极化与MIMO阵列雷达的角度估计技术

主　编：郑桂妹
副主编：宋玉伟　胡国平

西北工业大学出版社
西　安

【内容简介】 本书围绕极化分集的极化敏感阵列和波形分集的MIMO阵列的目标角度(包括低仰角目标的俯仰角)和极化信息测量问题展开，主要内容包括小电偶极子和小半径磁环的极化敏感阵列阻抗问题、辐射效率低的参数估计问题、MIMO阵列相干信号的解相干参数估计问题、米波MIMO阵列雷达导向矢量相互渗透的测高问题、极化分集和波形分集相结合的米波极化MIMO阵列雷达的测高问题等。

本书可用作从事阵列信号处理、雷达与通信系统、信号和信息处理、微波和电磁场、水声等专业高年级本科生和研究生教材，也可供相关技术领域研究人员及工程技术人员阅读参考。

图书在版编目(CIP)数据

极化与MIMO阵列雷达的角度估计技术/郑桂姝主编.—西安:西北工业大学出版社,2023.1

ISBN 978-7-5612-8645-6

Ⅰ.①极… Ⅱ.①郑… Ⅲ.①阵列雷达-研究 Ⅳ.①TN959

中国国家版本馆CIP数据核字(2023)第029479号

JIHUA YU MIMO ZHENLIE LEIDA DE JIAODU GUJI JISHU

极化与MIMO阵列雷达的角度估计技术

郑桂姝 主编

责任编辑: 王梦妮　　**策划编辑:** 杨　军
责任校对: 曹　江　　**装帧设计:** 李　飞
出版发行: 西北工业大学出版社
通信地址: 西安市友谊西路127号　　邮编:710072
电　　话: (029)88491757，88493844
网　　址: www.nwpup.com
印 刷 者: 陕西向阳印务有限公司
开　　本: 727 mm×960 mm　　1/16
印　　张: 9
字　　数: 166千字
版　　次: 2023年1月第1版　　2023年1月第1次印刷
书　　号: ISBN 978-7-5612-8645-6
定　　价: 49.00元

前　　言

参数估计是阵列雷达或通信系统重点关注的指标之一。常规阵列信号的参数估计算法研究已涌现出很多经典超分辨算法，而分集技术的发展使参数估计效果更加理想。目前，针对极化分集的极化敏感阵列雷达和波形分集的MIMO雷达的参数估计已有一定的研究成果，但仍存在一些关键问题需要深入研究。鉴于此，本书从以下几个方面开展研究，研究成果可以为基于分集技术的阵列信号参数估计问题提供更为扎实的理论支撑。

（1）针对极化分集的短电偶极子和小磁环存在辐射效率低下的问题，利用长电偶极子和大磁环组成的阵列来提高辐射效率，并提出一种基于ESPRIT的算法来估计目标的两维角度。

（2）针对波形分集的FDA－MIMO雷达在角度和距离联合估计过程中出现的距离模糊问题，提出利用参差频率增量来扩大距离模糊范围，并采用ESPRIT和MUSIC联合算法来降低计算量。此外，针对波形分集的速度传感器MIMO阵列的相干源测向，提出一种新的速度场平滑算法来解相干源。

（3）对波形分集的米波MIMO雷达的测高问题进行研究。

（4）鉴于极化分集和波形分集的优势，将两种优势相结合起来研究米波极化MIMO雷达的测高问题。

本书得到了西安电子科技大学雷达信号处理国家级重点实验室陈伯孝教授、杨明磊教授，清华大学电子系汤俊教授，海南大学王咸鹏教授，国防科技大学师俊朋副教授、刘章孟研究员，大连理工大学万良田副教授，空军工程大学张群教授的鼓励、帮助和支持，也得到了西北工业大学出版社的慷慨帮助，编辑为本书花了大量的精力。写作本书曾考阅了相关文献、资料，谨此一并表示由衷的感谢。

由于笔者水平有限，书中难免会有不妥之处，敬请同行与读者批评指正。

郑桂妹

2022年8月22日

目　　录

第1章 绪　　论

§1.1　研究背景和意义

从第二次世界大战至今的多数战争来看,雷达作为战争中的“千里眼”发挥着举足轻重的作用,是决定战争胜负的关键因素。阵列雷达是雷达发展至今的一个重要分支,其中,参数估计问题是阵列信号处理的核心问题之一。阵列参数估计是利用阵列接收的多通道数据进行处理从而得到目标的参数,如目标的角度和极化信息等。目前阵列的参数估计已经取得了丰硕的研究成果,但是随着现代战争中电子干扰的复杂多变,阵列信号参数估计的电磁环境严重恶化,对阵列信号参数估计的要求越来越高,这些变化敦促研究人员不断地进行技术革新,研究新的阵列体制和相应的参数估计理论与方法。

基于极化分集的极化敏感阵列是在这种背景下应运而生的一种新体制阵列雷达,其中一个重要特点是用矢量天线取代传统的标量天线,其阵列天线相较于常规阵列雷达有所不同。矢量天线由摆放方向不同的天线组成,因此能够感知极化信息,将这种以天线形式组成的阵列称为极化敏感阵列。一个完备的矢量天线由三个正交电偶极子和三个正交磁环组成。在这六个元素中,可以根据实际需求取任意两个以上的元素构成矢量天线。极化敏感阵列能够获取信号的极化分集信息。与常规阵列相比,极化敏感阵列在参数估计方面具有诸多优势:在受限空间中能够获取更多的通道信息,从而得到更高的估计精度;能够获取目标极化信息,更有利于目标特征的估计识别等。

与传统相控阵雷达相比,MIMO 雷达具有波形分集和空间分集的优势,是近年来提出的一种新体制雷达。本书主要研究只具有波形分集的集中式 MIMO 雷达,其虚拟孔径扩展的优势体现得更加明显,参数估计所能得到的优势更大。大虚拟孔径带来的良好参数估计性能吸引了大批学者和工业部门进行研究,取得了丰硕的成果,如多种超分辨算法的应用,从估计精度、算法计算量、工程可实现性等方面取得了较大突破,但是针对 MIMO 雷达相干、低仰角多径目标的角度与高度测量仍处于起步阶段。

极化分集的极化敏感阵列雷达系统和波形分集的 MIMO 阵列雷达的出

现是雷达发展过程的重要创新，分集技术在参数估计中体现出的重要优势受到了研究人员和工业部门的重点关注。基于极化分集的极化敏感阵列和波形分集的 MIMO 阵列参数估计已经取得了较为丰硕的成果，但其理论研究仍需深入，距离装备应用仍有一段距离，还存在一些关键技术需要突破。例如：传统短电偶极子和小半径磁环构成的极化矢量阻抗小、辐射效率低的问题；MIMO 阵列解相干的问题，尤其是米波 MIMO 阵列雷达导向矢量相互渗透的测高问题；以及如何更好地将极化分集和波形分集相结合的米波极化 MIMO 雷达测高问题；等等。即基于极化分集和波形分集的阵列信号参数估计仍是值得深入研究的课题，这些问题的解决将为分集技术的装备应用提供理论依据和技术支撑。

§1.2 国内外研究现状

§1.2.1 极化分集的极化敏感阵列参数估计

极化敏感阵列能够获得极化分集，而极化分集能够在滤波、检测等多个方面带来诸多好处，尤其是目标参数估计方面，故基于极化分集的极化敏感阵列目标参数估计问题受到了学者和工业部门的广泛关注，下面首先列出国内外的主要研究者和研究团队。国外主要有美国斯坦福大学 Ferrara 博士和 Parks 博士[1]，美国俄亥俄州立大学 Jian Li 博士[2-5]，以色列特拉维夫大学 Weiss 博士和美国加利福尼亚大学 Friedlander 教授[6-11]，美国伊利诺伊州立大学 Nehorai 教授团队[12-24]，以色列特拉维夫大学 Tabrikian 教授[25-26]，法国的 Bihan 博士[27-29]等；国内主要有北京航空航天大学黄启南（Kainam Thomas Wong）教授和美国贝尔实验室袁鑫（Xin Yuan）博士的团队[30-47]，国防科技大学徐振海、庄钊文、肖顺平等教授的团队[48-53]，北京理工大学徐友根、刘志文教授的团队[54-64]，大连理工大学龚晓峰教授的团队[65-72]，上海交通大学何劲研究员的团队[73-76]，复旦大学张建秋教授的团队[77-79]，西安电子科技大学王兰美和廖桂生[80-83]、杨明磊[84]教授的团队，南京航空航天大学张小飞教授的团队[85-88]，空军航空大学陶建武教授的团队[89-92]，电子科技大学王建英博士[93-95]，哈尔滨工程大学司伟建研究员的团队[96-100]，空军工程大学郑桂妹副教授[101-114]，等等。

针对极化敏感阵列的参数估计：若按天线结构来分的话，最典型的可分为共点式矢量天线极化敏感阵列和分离式极化敏感阵列；若按单个矢量天线的阵元数量来分，可分为两分量交叉偶极子矢量阵列的参数估计、三分量正交偶极子或磁环的极化矢量天线阵列、四分量极化矢量天线、五分量极化矢量天线、完整六分量电磁矢量传感器天线；若按参数估计算法可分为矢量叉积算法、经典超分辨算法、稀疏表征和压缩感知或者其他人工智能的方法，其中超分辨算法包括 ESPRIT、MUSIC、最大似然、子空间拟合等算法，人工智能的方法包括神经网络、BOMP 等算法，以及几种算法的组合应用；按照阵列结构的视角来分，可分为线阵、二维平面阵和二维圆阵、L 形阵列以及三维结构阵列等，其中重点有单个分离式矢量天线的结构优化布阵设计和整体阵列的结构排列设计；从噪声的角度来看，主要研究有高斯白噪声、色噪声以及阵列接收非均匀噪声；从信号源的视角来看，可分为已知信号波形形式和信号形式服从高斯分布的未知信号。阵列结构、信号形式、噪声形式以及相应的参数估计算法之间的配合主要取决于实际应用场景。有的算法和阵列结构是解决前人所留下的未解决的问题，更多的算法和阵列结构的形式是由实际应用场景来决定的，各种算法之间并没有绝对的好坏之分。同一种算法在此场景优于另一种算法，换一个应用场景则未必是相同的结果。下面挑选一小部分重要工作进展将极化敏感阵列的参数估计发展历史串联起来。

首先是美国斯坦福大学 Ferrara 教授等利用交叉偶极子组成的极化敏感阵列解决了 DOA 估计问题[1]，接着美国俄亥俄州立大学 Jian Li 教授等采用经典的 ESPRIT 超分辨算法解决了交叉偶极子组成的极化敏感阵列[2]、极化均匀加权平滑降低计算复杂度[4]、多分量相位比值法来实现参数估计等问题[5]。同一时期的交叉偶极子多项式求根[9]、最大似然[11]、模拟退火算法、子空间拟合等。Nehorai 教授在 1994 年提出完整电磁矢量传感器的概念来进行参数估计[12]，随后解决了部分阵列形式的线性独立性问题并解决了目标 DOA 估计的唯一性问题[13-16]。紧接着文献[22－24]中提出了一个电磁矢量传感器空间分离的阵列结构，以减少电磁互耦的影响，并说明了其在 DOA 估计中的优势所在。Wong 教授团队在 1997 年提出一种非常经典的矢量叉积算法[30]，并利用此算法作为粗估计参考来扩展阵列孔径，从而提供参数估计精度[32-34]。

接着，法国的 Bihan 博士等将多元代数推广至极化敏感阵列[27]，针对电磁矢量传感器 DOA 估计，给出了四元数[28]和双四元数[29] MUSIC 超分辨算法。在同一时期，国内在北京理工大学徐友根教授的带领下，龚晓峰副教授挖掘了基于张量和多元代数的极化敏感阵列 DOA 估计的优势[69-71]。2004 年，针对相干源入射信号的情况，Rahamim 博士巧妙地提出用极化平滑算法来代替空间平滑算法[26]，此算法并不损失阵列的有效孔径，在空间受限场景下应用效果更佳。龚晓峰教授和何劲教授对其进行推广，提出复数域极化平滑[71]和极化差分平滑[74]等经典算法。此外，在 2011 年，Wong 教授团队创造性地优化了单个分离式电磁矢量传感器的六个元素的摆放间距，使经典矢量叉积算法得以恢复应用[43-44]。2017 年，由 Wong 教授团队提出的以辐射效率更高的长电偶极子和大磁环代替短点偶极子和小磁环，来进行阵列流形、角度和极化参数估计算法的研究[45-47]。

在上述极化敏感阵列参数估计发展过程中，多位学者进行补充研究来完善其理论，包括：徐振海教授的极化域-空域联合滤波、检测、估计算法[48-52]；徐友根教授的二维[55]、三维[56]、四维[57]、五维矢量天线阵列的角度和极化参数联合估计方法中的导向矢量的秩模糊问题；龚晓峰教授的平行因子分析[68]、双四元数[69]、四-四元数[67]等算法；何劲研究员的传播算法[73]、色噪声场景[75]等算法；张建秋教授团队建立了三维空间几何代数在极化敏感阵列 DOA 估计的信号模型[77]；张小飞教授等解决了极化敏感阵列的盲 DOA 和极化参数的联合估计问题[87,88]；陶建武教授研究了近场源的 DOA、目标距离和极化参数等多维参数估计[90,92]；王建英博士等研究了在窄带、宽带条件下，目标载频、DOA 和极化参数的多维联合估计[94,95]；司建伟团队进行了降维联合参数估计研究[96,99]；郑桂妹副教授团队则在稀疏分离式电磁矢量阵列的参数估计做出贡献[101-103]。此外，李槟槟博士等则对大电偶极子和大磁环的高辐射效率参数估计方面提出矢量叉积、ESPRIT 算法等经典算法的移植应用[109-111]。

§1.2.2 波形分集的 MIMO 雷达参数估计

波形分集的 MIMO 雷达时代[115-122]从 2004 年开始到来，至今已发展十几年。MIMO 雷达按照天线的布阵距离分类：一种是同时具有波形分集和空间分集的统计式 MIMO 雷达；另一种是阵列天线距离很近，波长级别的，只具有波形分集的集中式 MIMO 雷达。这两种方式的 MIMO 雷达在目标检测、参数估计、识别成像等重要环节具有优势，其中，很重要的一个环节是集中式 MIMO 雷达的参数估计，尤其是目标 DOA 角度估计。因此，本书重点研究基于波形分集 MIMO 雷达的 DOA 角度等参数估计。目前国内外研究 MIMO 雷达的学者和团队众多，一一列举较为不便，故下面主要以研究单位的形式来总结发展现状。国外开展基于波形分集的 MIMO 雷达目标参数估计的国家主要有美国[123-127]、以色列[128]、法国[129,130]、芬兰[131]、韩国[132]等。国内开展相关 MIMO 雷达参数估计的单位主要有国防科技大学[133-135]、北京理工大学[136-137]、电子科技大学[138-139]、西安电子科技大学[140-159]、南京航空航天大学[160-169]、南京理工大学[170-174]、哈尔滨工程大学[175-188]、深圳大学[189-190]、海南大学[191-201]、空军工程大学[202-217]以及其他多家单位[218-220]。

MIMO 雷达参数估计的研究核心总结为以下几个方面：一是提高参数估计精度的算法层面，主要包括经典超分辨算法的移植和改进与基于压缩感知和人工智能新算法的开发，包括 ESPRIT、MUSIC、最大似然、BOMP、张量技术等；二是由于 MIMO 雷达匹配滤波后通道增多，自由度成倍地增加导致算法的计算量激增，故降低计算量的算法亦成为研究热点，包括降维 Capon、降维 MUSIC、传播算子等降低计算量的方法；三是双基地 MIMO 雷达和多维参数估计的多目标参数之间的配对问题，包括利用实值处理、MIMO 数据镶嵌特性等；四是米波 MIMO 雷达目标测高问题研究，如广义 MUSIC 算法、光滑地面、起伏地面、导向矢量相互渗透等问题；五是 MIMO 阵列雷达阵列优化设计问题，包括均匀稀疏和非均匀稀疏，如嵌套阵和互质阵，可归结为同等阵元数最大自由度的挖掘；六是从信号形式和噪声角度来看，信号可包括圆信号和非圆信号，噪声可分为高斯白噪声、色噪声、冲击噪声、非均匀噪声等；七是非理想条件下的 MIMO 雷达参数估计问题，包括阵元间互耦、阵元位置误差、非理想正交发射波形等；八是结合载频分集阵列雷达的 FDA-MIMO 雷达的参数估计问题。下面对这几个方面进行详细阐述。

第一方面是关于经典算法的应用，ESPRIT 是利用空域转不变空间的原理来估计参数，对于发射正交波形的 MIMO 雷达来说，发射信号匹配之后等距线阵的发射导向矢量和接收导向矢量均具有旋转不变性，故文献[140]、[144]、[170]、[207]等分别对双基地 MIMO 雷达应用旋转不变子空间算法得到发射角和接收角的联合估计，并利用 MIMO 雷达特有的矩阵结构和实值处理来实现多目标的自动配对，且用共轭阵列的形式来提供非圆信号的估计精度。文献[126]、[127]对发射的非正交信号进行波形优化设计，使其在信噪比最大的前提下依然能够保持空间旋转不变性。文献[152]、[161]对单基地 MIMO 雷达的 DOA 估计进行整体的维度降解，以减少计算量，最后采用无须搜索的 ESPRIT 算法求得目标 DOA。针对传统 MUSIC 算法，由于其具有二维参数，需要的二维搜索计算量很大，故文献[129]、[145]对其进行降维处理，并利用求根算法来避免搜索以降低计算量。此外还有一些利用 ESPRIT 与 MUSIC 相结合的办法来联合估计发射角和接收角[130-149]，还有最大似然算法的应用[139]。对于稀疏表征和压缩感知的 MIMO 雷达角度估计，主要有 l_0 范数[177]、协方差矢量稀疏表征[178-179]、稀疏贝叶斯学习[218]。还有开发关于张量数据模型的色噪声和互耦条件下的 MIMO 雷达角度估计[181-182]。第二方面是关于降低计算量的方法，主要有降维、非搜索类算法，包括 ESPRIT 和求根 MUSIC，波束域求根 MUSIC[211] 等。第三方面结合第一方面的内容已经阐述。关于第四方面测高问题，文献[146]利用最大似然方法对四条接收导向矢量相互渗透的情况进行处理，得到测高结果。文献[158]利用稀疏字典收缩的方法得到复杂地形下的测高结果，效果良好。第五方面是关于 MIMO 阵列雷达的阵列优化设计，主要有嵌套 MIMO 阵列的张量模型及其角度估计算法[134]，互质阵 MIMO 雷达的二维角度估计[217]。第六方面是关于信号和噪声形式的研究，主要有空间色噪声以及未知相关噪声的参数估计算法[154,155,220]、圆信号和非圆信号[187,192]、色噪声[199]以及相干分布式信号的中心角度估计[204]。第七方面是关于非理想状态下的参数估计算法，如非理想正交波形的白化和快速算法[189,209,219]，以及互耦和阵列校准的参数估计算法[167,182,195,203]。第八方面是关于 FDA-MIMO 雷达参数估计问题的研究[221-226]。FDA 在雷达的很多应用方面都具有其独特优势[221-222]。它的方向图不再是传统的 DOA 角度的函数，而是依赖于目标距离的函数，若不做特殊处理，则角度和距离是耦合的。因此，它在目标抗干扰等方面有独特的优势。MIMO 雷达由于具有传统相控阵雷达没有的波形分集优势，故 FDA-MIMO 雷达受到了广泛的关注[223]。其中对于角度和距离的联合估计是 FDA-MIMO 雷达的一个重要研究课题[224-226]。

§1.2.3 极化分集与波形分集的极化 MIMO 雷达参数估计

关于极化分集和波形分集相结合的极化 MIMO 雷达参数估计的研究，国外的主要研究学者有 Chintagunta 博士[227] 和 Bencheikh 博士[228] 等，国内主要有吉林大学的姜虹教授团队[229-230]、南京理工大学的何劲教授团队[231-233] 与顾红教授团队[234]、长江大学的文方青副教授团队[235-236]、海南大学的王咸鹏团队[237]、西安电子科技大学的马慧慧博士[238] 以及空军工程大学的郑桂妹副教授与西安电子科技大学等人合作的团队[239-243] 等。极化 MIMO 雷达的发展是随着分离式极化矢量和 MIMO 技术的发展而发展的，从经典的共点式极化矢量发展到分离式极化矢量 MIMO 的优化设计，算法从 ESPRIT 超分辨算法到平行因子分析以及块正交匹配追踪的稀疏表征算法，紧接着是极化平滑算法等在极化 MIMO 雷达中的应用。以上研究均证明极化分集和波形分集相结合能够提升雷达系统的性能。

综上所述，极化分集的极化敏感阵列和波形分集的 MIMO 阵列参数估计问题主要集中在提高极化敏感阵列的辐射效率、参数估计精度的提升、相干信号的参数估计问题，尤其是 MIMO 雷达的测高问题。本书也围绕这几方面进行深入研究。

§1.3　本书的主要内容安排

本书围绕极化分集的极化敏感阵列和波形分集的 MIMO 阵列的目标角度(包括低仰角目标的俯仰角)与极化信息测量问题展开，主要解决如下几个关键技术问题，如：小电偶极子和小半径磁环的极化敏感阵列阻抗小、辐射效率低的参数估计问题；MIMO 阵列相干信号的解相干参数估计问题；米波 MIMO 阵列雷达导向矢量相互渗透的测高问题；极化分集和波形分集相结合的米波极化 MIMO 阵列雷达的测高问题等。

本书的总体内容结构框架如图 1.1 所示，全书内容安排如下：

第 1 章为绪论，首先介绍本书的研究背景和意义，然后对基于极化分集和波形分集的参数估计算法的国内外研究现状作阐述与剖析，最后对本书的工作和各章节研究内容做出安排。

第 2 章阐述基于极化分集技术的极化敏感阵列、基于波形分集技术的 MIMO 雷达参数估计的信号模型和经典角度估计算法，首先给出极化敏感阵列的基本知识，然后对阵列构型进行分析，紧接着是信号模型的推导，并给出相对应的经典 MUSIC 超分辨算法，最后给出计算机仿真结果，验证极化敏感阵列的参数估计。另外一部分内容是针对波形分集的 MIMO 雷达，做了与极化敏感阵列类似的布局和分析。

第 3 章针对极化分集的短电偶极子和小半径磁环存在辐射效率低的问题，利用长电偶极子和大半径磁环组成的阵列来提高辐射效率，并提出一种基于 ESPRIT 的算法来估计目标的二维角度。该阵列和所提参数估计算法有如下优点：阵元辐射效率高、算法无需搜索计算量小、分离式矢量天线阵元间互耦小。成果可为基于极化敏感阵列的雷达或通信系统的工程应用提供强有力的理论支持。

第 4 章研究 MIMO 雷达角度与距离参数估计问题。针对波形分集的 FDA-MIMO 雷达在角度和距离联合估计过程中出现的距离模糊问题，利用参差脉冲解模糊的思想，利用发射参差频率增量来提高出现栅瓣的距离，并提出一种 ESPRIT - MUSIC 算法得到无模糊距离和角度估计值，其中角度和距离是自动配对的。此外,对波形分集的速度传感器 MIMO 阵列进行相干源测向,提出一种新的速度场平滑算法来解相干源,并分析其去相关性能。所提算法不需要速度矢量传感器的位置信息,适用于任意构型阵列,且算法没有损失阵列有效孔径。

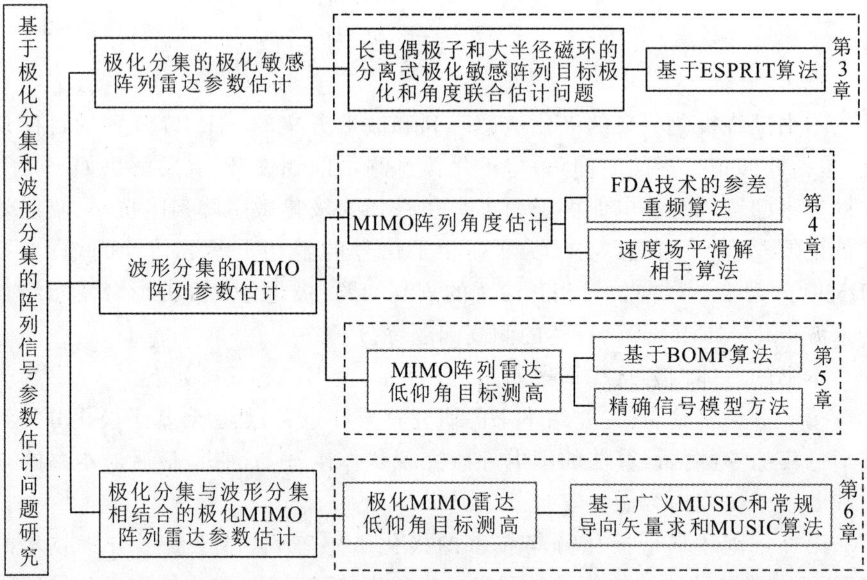

图 1.1　本书总体结构框架

第 5 章研究米波 MIMO 雷达的目标高度参数估计问题,即测高问题。最大似然和广义 MUSIC 方法是米波 MIMO 阵列雷达测高方法行之有效的算法,但其计算量大,工程中难以接受。为此,本书提出一种基于 BOMP 预处理的方法来降低计算量,主要利用粗栅格得到角度粗估计,再以此角度粗估计为初始值中心,取 MIMO 雷达波束宽度作为搜索范围来进行搜索。此外,针对米波 MIMO 雷达的复杂起伏阵地,建立精确反射信号模型,然后利用最大似然和广义 MUSIC 方法来验证精确信号模型下即地形匹配算法的测高效果,仿真结果表明所提地形匹配算法比地形失配的算法估计精度好得多。

第 6 章鉴于极化分集和波形分集的优势,将两种优势相结合,研究极化 MIMO 雷达的测高问题。首先推导两种分集技术相结合的平坦镜面反射的测高模型。然后对信号模型进行适当的归类和变形,使其能够适用于经典超分辨 MUSIC 测高算法,包括广义 MUSIC 算法和导向矢量合成 MUSIC 算法,并给出两种算法的计算复杂度和极化 MIMO 雷达测高的 CRB 推导结果。

推导过程表明研究方法并不需要解相干处理，这是所提算法的一大优势。计算机仿真结果验证了所提测高算法的有效性。

第 7 章对本书所做工作进行梳理和总结，以及对未来的研究方向进行展望。

第2章　基于极化分集技术的极化敏感阵列与波形分集技术的MIMO阵列参数估计基础

§2.1　极化分集技术的极化敏感阵列参数估计基础

§2.1.1　极化敏感阵列基本知识[49][54]

1.电磁波传播极化特性

电磁波传播有两个重要的特性，空间信息和极化信息。在图2.1所示的直角坐标系中，采用俯仰角和方位角来表示电磁波的空间信息，用θ表示俯仰角，用ϕ表示方位角。电磁波传播的两个垂直分量的大小关系，可以用极化辅角γ和极化相位差η来表示。当然这里针对的是完全极化波，部分极化波不是本书的研究内容，这里不作展开讨论。

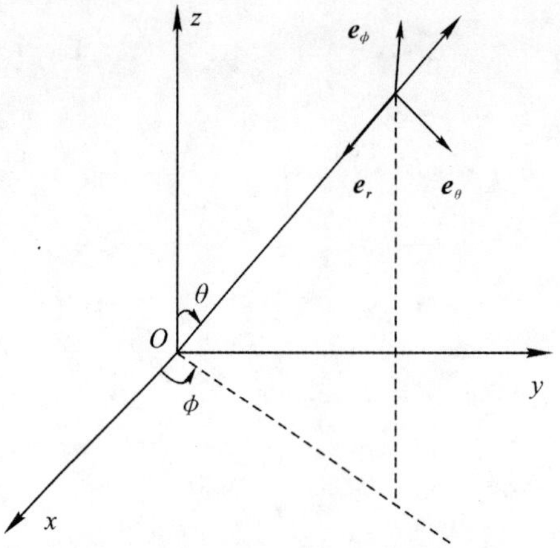

图2.1　电磁信号传播示意图

假设电磁波是远场平面波入射。那么电磁波入射到原点阵元的接收电场可表示为

$$\boldsymbol{E}=E_{\phi}\boldsymbol{e}_{\phi}+E_{\theta}\boldsymbol{e}_{\theta} \tag{2.1}$$

其中，$\boldsymbol{e}_r$，$\boldsymbol{e}_\phi$，$\boldsymbol{e}_\theta$是满足右手螺旋法则的单位传播矢量。E_ϕ 为 $\boldsymbol{e}_\phi$ 方向的电场大小，E_θ 为 $\boldsymbol{e}_\theta$方向的电场大小，E_ϕ 和 E_θ 之间的大小关系，简称为极化信息，且有如下的关系式：

$$\begin{bmatrix} E_\theta \\ E_\phi \end{bmatrix} = \begin{bmatrix} \cos\gamma \\ \sin\gamma \mathrm{e}^{\mathrm{j}\eta} \end{bmatrix} \tag{2.2}$$

为了表示方便，将$(\boldsymbol{e}_r,\boldsymbol{e}_\theta,\boldsymbol{e}_\phi)$三个方向转化到$(\boldsymbol{e}_x,\boldsymbol{e}_y,\boldsymbol{e}_z)$直角坐标系中，并将电场强度用直角坐标系中的三个量表示为

$$[\boldsymbol{e}_r\ \boldsymbol{e}_\theta\ \boldsymbol{e}_\phi] = [\boldsymbol{e}_x\ \boldsymbol{e}_y\ \boldsymbol{e}_z] \begin{bmatrix} \cos\phi\sin\theta & \cos\phi\cos\theta & -\sin\phi \\ \sin\phi\sin\theta & \sin\phi\cos\theta & \cos\phi \\ \cos\theta & -\sin\theta & 0 \end{bmatrix} \tag{2.3}$$

$$\boldsymbol{E} = [\boldsymbol{e}_x\ \boldsymbol{e}_y\ \boldsymbol{e}_z] \begin{bmatrix} \sin\gamma \mathrm{e}^{\mathrm{j}\eta}\cos\varphi\cos\theta - \cos\gamma\sin\varphi \\ \sin\gamma \mathrm{e}^{\mathrm{j}\eta}\sin\varphi\cos\theta + \cos\gamma\cos\varphi \\ -\sin\gamma \mathrm{e}^{\mathrm{j}\eta}\sin\theta \end{bmatrix} \tag{2.4}$$

根据右手螺旋定理可计算得到磁场强度：

$$\begin{aligned} \boldsymbol{H} &= -(\boldsymbol{e}_r \times \boldsymbol{E}) \\ &= -[\boldsymbol{e}_r \times (E_\phi \boldsymbol{e}_\phi + E_\theta \boldsymbol{e}_\theta)] \\ &= -E_\phi(\boldsymbol{e}_r \times \boldsymbol{e}_\phi) - E_\theta(\boldsymbol{e}_r \times \boldsymbol{e}_\theta) \\ &= E_\theta \boldsymbol{e}_\phi - E_\phi \boldsymbol{e}_\theta \end{aligned} \tag{2.5}$$

同理利用直角坐标系中的转换关系可以得到磁场矢量：

$$\boldsymbol{H} = [\boldsymbol{e}_x\ \boldsymbol{e}_y\ \boldsymbol{e}_z] \begin{bmatrix} -(\cos\gamma\cos\varphi\cos\theta + \sin\gamma \mathrm{e}^{\mathrm{j}\eta}\sin\varphi) \\ \cos\gamma\sin\varphi\cos\theta - \sin\gamma \mathrm{e}^{\mathrm{j}\eta}\cos\varphi \\ \cos\gamma\sin\theta \end{bmatrix} \tag{2.6}$$

由式(2.4)和式(2.6)知，电磁波的完备电场和磁场矢量为

$$\begin{bmatrix} \boldsymbol{E} \\ \boldsymbol{H} \end{bmatrix} = \begin{bmatrix} -\sin\phi & \cos\theta\cos\phi \\ \cos\phi & \cos\theta\sin\phi \\ 0 & -\sin\theta \\ -\cos\theta\cos\phi & -\sin\phi \\ \cos\theta\sin\phi & -\cos\phi \\ \sin\theta & 0 \end{bmatrix} \begin{bmatrix} \cos\gamma \\ \sin\gamma \mathrm{e}^{\mathrm{j}\eta} \end{bmatrix} \tag{2.7}$$

2.极化敏感阵列常用结构

利用电偶极子和磁环在空间上排列构成一定的阵列结构，我们把这种阵列称为极化敏感阵列。极化敏感阵列的布置与常规阵列大体一致，主要有均

匀线阵、均匀圆阵、均匀矩形阵列结构、L 形阵列结构。其中，除了第一种是一维阵列结构，只能获取一维角度信息，其他三种均为二维阵列结构，可以获得二维角度估计。从上面的分析可以得到，要想获取极化信息，至少需要知道式(2.7)中的两个分量，故极化敏感阵列中的极化天线至少由电偶极子和磁环中的两个量组成。根据分量的多少，在极化敏感阵列的参数领域中，把两个电偶极子组成的极化天线称为交叉偶极子，如图 2.2 所示，把三个电偶极子组成的极化矢量天线称为三正交电偶极子，如图 2.3 所示，将三个相互正交的电偶极子和三个相互正交的磁环空间上共点放置称为电磁矢量传感器，如图 2.4 所示。需要注意的是，目前的极化敏感阵列均相互垂直摆放，或者沿着直角坐标系三个轴摆放。

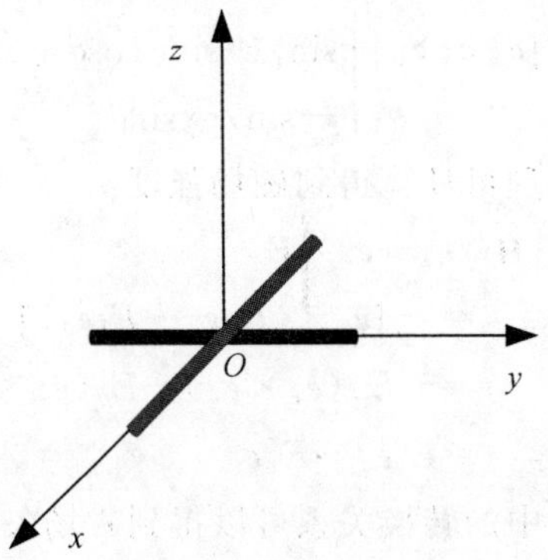

图 2.2　交叉偶极子构成的极化矢量天线[114]

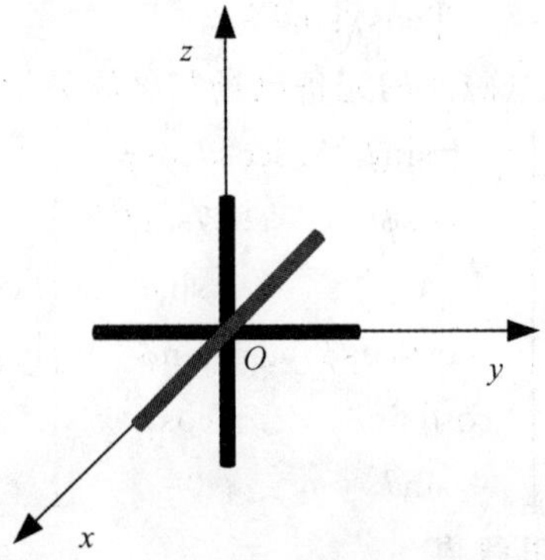

图 2.3　三正交电偶极子构成的极化矢量天线[114]

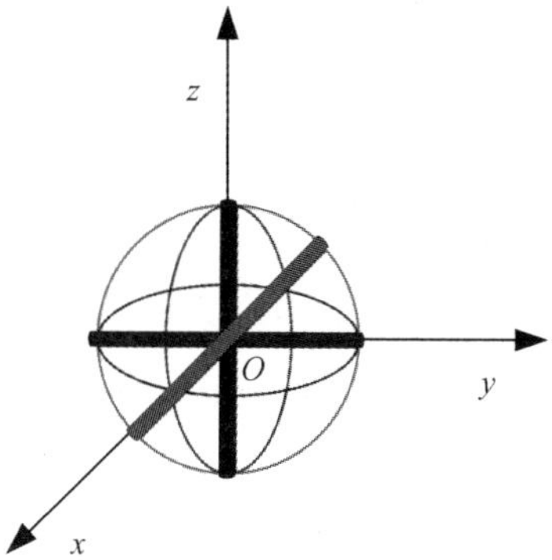

图 2.4　三正交电偶极子和磁环共点放置的电磁矢量传感器[114]

§ 2.1.2　极化敏感阵列信号接收模型

依据前两节阐述的极化敏感阵列知识，以交叉偶极子极化敏感阵列为例来说明其参数估计问题，如图 2.5 所示。

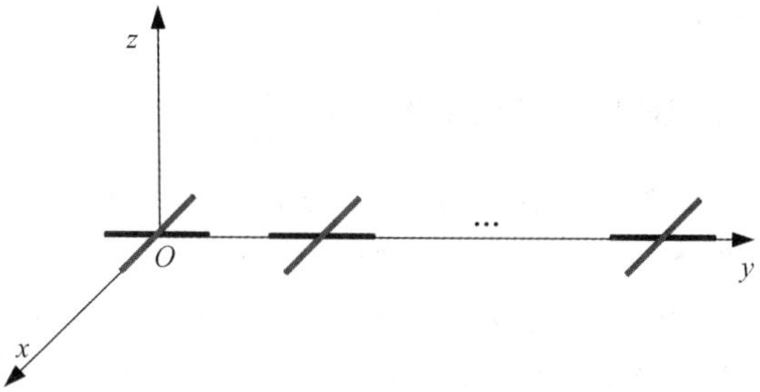

图 2.5　交叉电偶极子极化敏感阵列

本节建立由正交偶极子构成的均匀线阵信号接收模型，其他阵列形式和矢量天线形式以此类推。

由式(2.7)可以得到，单个矢量天线的极化导向矢量为取前两个分量，即

$$\boldsymbol{\alpha}_p(\theta,\phi,\gamma,\eta)=\begin{bmatrix}-\sin\phi\cos\theta\cos\phi\\ \cos\phi\cos\theta\sin\phi\end{bmatrix}\begin{bmatrix}\cos\gamma\\ \sin\gamma\mathrm{e}^{\mathrm{j}\eta}\end{bmatrix}\tag{2.8}$$

众所周知，均匀线阵并没有二维角度的能力，不失一般性，可设置方位角 $\phi=90°$，则极化导向矢量转化为

$$\boldsymbol{\alpha}_p(\theta,\gamma,\eta)=\begin{bmatrix}-\cos\gamma\\ \cos\theta\sin\gamma\mathrm{e}^{\mathrm{j}\eta}\end{bmatrix}\tag{2.9}$$

假设阵列由 M 个阵元组成，相邻阵元间距为 d，则阵列的空域导向矢量为

$$\boldsymbol{\alpha}_s(\theta)=\begin{bmatrix}1\\ e^{-j2\pi d\sin\theta/\lambda}\\ \cdots\\ e^{-j2\pi d\sin\theta/\lambda(M-1)}\end{bmatrix} \tag{2.10}$$

其中，λ 表示入射波长。则整个极化敏感阵列的极化域-空域联合导向矢量可表示为

$$\boldsymbol{\alpha}(\theta,\gamma,\eta)=\boldsymbol{\alpha}_p(\theta,\gamma,\eta)\otimes\boldsymbol{\alpha}_s(\theta) \tag{2.11}$$

单个目标入射到该阵列，其接收信号模型可表示为

$$\boldsymbol{x}(t)=\boldsymbol{\alpha}(\theta,\gamma,\eta)s(t)+\boldsymbol{n}(t) \tag{2.12}$$

其中，$s(t)$为入射信号，$\boldsymbol{n}(t)$为噪声矢量。若有 K 个信号入射到该阵列上，将式(2.12)转化为如下的矩阵形式：

$$\boldsymbol{x}(t)=\boldsymbol{A}\boldsymbol{s}(t)+\boldsymbol{n}(t) \tag{2.13}$$

其中，$\boldsymbol{s}(t)=[s_1(t)\ s_2(t)\ \cdots\ s_K(t)]$为接收信号矢量，$\boldsymbol{A}$ 为导向矩阵，表示为

$$\boldsymbol{A}=[\boldsymbol{\alpha}(\theta_1,\gamma_1,\eta_1)\ \boldsymbol{\alpha}(\theta_2,\gamma_2,\eta_2)\ \cdots\ \boldsymbol{\alpha}(\theta_K,\gamma_K,\eta_K)] \tag{2.14}$$

§2.1.3 极化敏感阵列经典参数估计算法

已有研究成果表明，在相同的孔径条件下，极化敏感阵列雷达的 DOA 估计精度要高于普通标量阵列雷达。从第 1 章的国内外研究现状可知，极化敏感阵列参数估计有很多算法，如 ESPRIT、MUSIC、稀疏表征、压缩感知等。下面介绍两种经典超分辨算法——MUSIC 和 ESPRIT 算法。

1.MUSIC 算法[85-86]

按如下公式展开求解阵列协方差矩阵，即

$$\begin{aligned}\boldsymbol{R}_x&=E\{\boldsymbol{x}(t)\boldsymbol{x}(t)^{\mathrm{H}}\}\\&=\boldsymbol{A}E\{\boldsymbol{s}(t)\boldsymbol{s}(t)^{\mathrm{H}}\}\boldsymbol{A}^{\mathrm{H}}+E\{\boldsymbol{n}(t)\boldsymbol{n}(t)^{\mathrm{H}}\}\\&=\boldsymbol{A}\boldsymbol{R}_s\boldsymbol{A}^{\mathrm{H}}+\boldsymbol{R}_n\end{aligned} \tag{2.15}$$

其中，$\boldsymbol{R}_s$ 和 $\boldsymbol{R}_n$ 被分别定义为信号协方差矩阵和噪声协方差矩阵。对矩阵 $\boldsymbol{R}_x$ 进行特征值分解可得

$$\boldsymbol{R}_x=\boldsymbol{U}\boldsymbol{\Lambda}\boldsymbol{U}^{\mathrm{H}}=\sum_{i=1}^{2M}\lambda_i\boldsymbol{u}_i\boldsymbol{u}_i^{\mathrm{H}}=\boldsymbol{U}_s\boldsymbol{\Sigma}_s\boldsymbol{U}_s^{\mathrm{H}}+\boldsymbol{U}_n\boldsymbol{\Sigma}_n\boldsymbol{U}_n^{\mathrm{H}} \tag{2.16}$$

其中，λ_i 表示特征分解得到的第 i 个特征值，$\boldsymbol{u}_i$ 是特征值 λ_i 所对应的特征向量，所有特征向量 $\boldsymbol{u}_i$ 组成矩阵 $\boldsymbol{U}$，所有特征值组成对角矩阵 $\boldsymbol{\Lambda}$，$\boldsymbol{\Lambda}=\mathrm{diag}[\lambda_1\ \lambda_2$

$\cdots \lambda_M \lambda_{M+1} \cdots \lambda_{2M}]$，理想条件下，其中最小的 $2M-K$ 个特征值都等于噪声强度。下面将矩阵 $\boldsymbol{U}$ 划分为两个矩阵，即信号子空间 $\boldsymbol{U}_s$ 和噪声子空间 $\boldsymbol{U}_n$。$\boldsymbol{U}_s=[\boldsymbol{u}_1\ \boldsymbol{u}_2 \cdots \boldsymbol{u}_K]$，$\boldsymbol{U}_n=[\boldsymbol{u}_{K+1}\ \boldsymbol{u}_{K+2} \cdots \boldsymbol{u}_{2M}]$。

已有研究成果证明导向矩阵 $\boldsymbol{A}$ 和信号子空间 $\boldsymbol{U}_s$ 扩张成的空间相同，并和噪声子空间正交，可用下式表示：

$$\mathrm{span}\{\boldsymbol{A}\}=\mathrm{span}\{\boldsymbol{U}_s\} \tag{2.17}$$

$$\mathrm{span}\{\boldsymbol{A}\} \perp \mathrm{span}\{\boldsymbol{U}_n\} \tag{2.18}$$

式(2.18)中的导向矩阵与噪声子空间正交，则每列的导向矢量 $\boldsymbol{\alpha}(\theta,\gamma,\eta)$ 也与噪声子空间 $\boldsymbol{U}_n$ 正交，即

$$\boldsymbol{U}_n^{\mathrm{H}}\boldsymbol{\alpha}(\theta,\gamma,\eta)=0 \tag{2.19}$$

在实际应用中，很难得到无限快拍，故用有限 L 个快拍的采样数据来计算协方差矩阵 $\boldsymbol{R}_x$ 的估计 $\hat{\boldsymbol{R}}_x$，表达式为

$$\hat{\boldsymbol{R}}_x=\frac{1}{L}\sum_{l=1}^{L}\boldsymbol{x}(t_l)\boldsymbol{x}(t_l)^{\mathrm{H}} \tag{2.20}$$

由式(2.16)可以求得 $\boldsymbol{U}_n$ 的估计 $\hat{\boldsymbol{U}}_n$。在噪声条件下，$\boldsymbol{\alpha}(\theta,\gamma,\eta)^{\mathrm{H}}\boldsymbol{U}_n$ 并不是严格等于零，而是接近于零，故构造如下搜索空间谱：

$$P_{\mathrm{music}}=\frac{1}{\boldsymbol{\alpha}(\theta,\gamma,\eta)^{\mathrm{H}}\hat{\boldsymbol{U}}_n\hat{\boldsymbol{U}}_n^{\mathrm{H}}\boldsymbol{\alpha}(\theta,\gamma,\eta)} \tag{2.21}$$

用(θ,γ,η)张成的三维参数对式(2.21)进行搜索，其极值点坐标值即为目标的参数估计值。

根据上述分析，总结极化敏感阵列雷达 MUSIC 超分辨算法步骤如下：

第一步：在有限快拍条件下，根据式(2.20)计算得到阵列协方差矩阵 $\hat{\boldsymbol{R}}_x$。

第二步：对 $\hat{\boldsymbol{R}}_x$ 采用特征值分解，按照式(2.16)的方法将其分解为信号子空间 $\hat{\boldsymbol{U}}_s$ 和噪声子空间 $\hat{\boldsymbol{U}}_n$。

第三步：利用噪声子空间 $\hat{\boldsymbol{U}}_n$ 和导向矢量 $\boldsymbol{\alpha}(\theta,\gamma,\eta)$ 构造式(2.21)的空间谱。

第四步：目标个数对应空间谱的峰值数，取出峰值所对应的坐标即为目标 DOA 和极化参数估计值。

2.ESPRIT 算法[85-86]

上一小节已经求得信号子空间 $\boldsymbol{U}_s$，下面就要找到关于 $\boldsymbol{U}_s$ 的旋转不变特性。取图 2.5 所示的极化敏感阵列左边$(M-1)$个交叉电偶极子天线和该

阵列的右边$(M-1)$个交叉电偶极子天线构成两个空域平移后相同的旋转不变阵列，对于第k个入射信号源，该均匀线阵具有如下空域旋转不变特性：

$$\boldsymbol{J}_{\mathrm{psa},2}\boldsymbol{\alpha}(\theta_k,\gamma_k,\eta_k)=\mathrm{e}^{-\mathrm{j}2\pi d\sin\theta_k/\lambda}\boldsymbol{J}_{\mathrm{psa},1}\boldsymbol{\alpha}(\theta_k,\gamma_k,\eta_k) \tag{2.22}$$

其中，$\boldsymbol{J}_{\mathrm{psa},1}$和$\boldsymbol{J}_{\mathrm{psa},2}$表示选择矩阵，$\boldsymbol{J}_{\mathrm{psa},1}=[\boldsymbol{I}_{M-1}\ \boldsymbol{O}_{(M-1)\times1}\ \boldsymbol{I}_{(M-1)}\ \boldsymbol{O}_{(M-1)\times1}]$，$\boldsymbol{J}_{\mathrm{psa},2}=[\boldsymbol{O}_{(M-1)\times1}\ \boldsymbol{I}_{M-1}\ \boldsymbol{O}_{(M-1)\times1}\ \boldsymbol{I}_{(M-1)}]$，$\boldsymbol{J}_{\mathrm{psa},1}$表示选取导向矢量$\boldsymbol{\alpha}(\theta_k,\gamma_k,\eta_k)$的前$(M-1)$个元素与间隔$M$个阵元之后的$(M-1)$个元素，$\boldsymbol{J}_{\mathrm{psa},2}$表示选取导向矢量$\boldsymbol{\alpha}(\theta_k,\gamma_k,\eta_k)$的第2个阵元开始的$(M-1)$个元素与间隔$M$个阵元之后的$(M-1)$个元素。考虑所有$K$个目标，则能够把式(2.22)转化成矩阵的表达形式：

$$\boldsymbol{J}_{\mathrm{psa},2}\boldsymbol{A}=\boldsymbol{J}_{\mathrm{psa},1}\boldsymbol{A}\boldsymbol{\Phi}_{\theta} \tag{2.23}$$

$$\boldsymbol{\Phi}_{\mathrm{psa},\theta}=\mathrm{diag}[\mathrm{e}^{-\mathrm{j}2\pi d\sin\theta_1/\lambda}\ \ \mathrm{e}^{-\mathrm{j}2\pi d\sin\theta_2/\lambda}\cdots\ \mathrm{e}^{-\mathrm{j}2\pi d\sin\theta_K/\lambda}] \tag{2.24}$$

其中，$\boldsymbol{\Phi}_{\mathrm{psa},\theta}$只含有未知量DOA的角度信息。把$\boldsymbol{U}_s=\boldsymbol{AT}$($\boldsymbol{T}$为非奇异矩阵)的几何关系式代入式(2.23)则可求得关于信号子空间$\boldsymbol{U}_s$的等式方程：

$$\boldsymbol{J}_{\mathrm{psa},2}\boldsymbol{U}_s=\boldsymbol{J}_{\mathrm{psa},1}\boldsymbol{U}_s\boldsymbol{\Psi}_{\mathrm{psa},\theta} \tag{2.25}$$

其中，$\boldsymbol{\Psi}_{\mathrm{psa},\theta}=\boldsymbol{T}^{-1}\boldsymbol{\Phi}_{\mathrm{psa},\theta}\boldsymbol{T}$。

噪声无时无刻都存在，此时式(2.25)变成近似相等，那么可利用总体最小二乘等方法得到$\boldsymbol{\Psi}_{\mathrm{psa},\theta}$的闭式解。则DOA角度估计结果可利用下式计算得到：

$$\hat{\theta}_k=\arcsin\frac{-\lambda\angle([\boldsymbol{\Phi}_{\theta}]_{kk})}{2\pi d},\quad k=1,\cdots,K \tag{2.26}$$

其中，矩阵$\boldsymbol{\Phi}_{\theta}$的对角元素$\{[\boldsymbol{\Phi}_{\theta}]_{kk},k=1,\cdots,K\}$为$\boldsymbol{\Psi}_{\theta}$特征分解之后的特征值。需要注意的是由于式(2.25)中的等式关系是由空间平移半波长得到的，故式(2.26)的DOA估计结果在全空域无模糊。

交叉电偶极子极化天线是同点放置，图2.5中平行于水平轴的所有阵元分量与平行于垂直轴点所有阵元分量空域上是完全相同的，其差距在于一个极化比值，故可利用这种极化域旋转不变性来估计极化信息。那么该极化比值定义为极化导向矢量中的第二个分量与第一个分量之间的比值，则有

$$\Lambda_p=\frac{\sin\gamma_p\cos\varphi_p\mathrm{e}^{\mathrm{j}\eta_p}}{-\cos\gamma_p} \tag{2.27}$$

可通过这样的两个选择矩阵：$\boldsymbol{J}_{\mathrm{polar},2}=\{\boldsymbol{I}_{2M}\otimes[0,1]\}$，$\boldsymbol{J}_{\mathrm{polar},1}=\{\boldsymbol{I}_{2M}\otimes[1,0]\}$，并依据推导DOA估计的方法得到旋转不变方程：

$$\boldsymbol{J}_{\mathrm{polar},2}\boldsymbol{U}_s=\boldsymbol{J}_{\mathrm{polar},1}\boldsymbol{U}_s\boldsymbol{\Psi}_{\mathrm{polar}} \tag{2.28}$$

其中，$\boldsymbol{\Psi}_{\text{polar}}=\boldsymbol{T}^{-1}\boldsymbol{\Phi}_{\text{polar}}\boldsymbol{T}$，$\boldsymbol{\Phi}_{\text{polar}}=\text{diag}[\Lambda_1\ \Lambda_2\cdots\ \Lambda_K]$。

噪声条件下，式(2.28)变成近似相等，那么可利用总体最小二乘法得到 $\boldsymbol{\Psi}_{\Lambda}$ 的闭式解。进一步利用特征分解求得极化旋转不变因子 $\Lambda_k(k=1,\cdots,K)$，那么二维极化信息可表示为

$$\left.\begin{aligned}\gamma_k&=\tan^{-1}\left|\frac{-\Lambda_k}{\cos\phi_k}\right|\\ \eta_k&=\arg(-\Lambda_k)\end{aligned}\right\}\tag{2.29}$$

§2.1.4　极化敏感阵列目标参数估计仿真结果

假设远场有两个不相干信号射向阵列，其俯仰角为 $\theta=\{30°,60°\}$，相对应的极化辅角为 $\gamma=\{30°,45°\}$，极化相位差为 $\eta=\{90°,90°\}$。阵列为半波长布阵的均匀线阵。交叉偶极子阵元数 $M=10$。

仿真一：MUSIC 搜索空间谱。

信噪比 SNR 为 20 dB，快拍数为 512。为了方便说明，MUSIC 算法只进行二维搜索，设置极化相位差为已知。图 2.6 给出其二维搜索 MUSIC 空间谱，从图 2.6 中可以看出 MUSIC 谱具有两个谱峰，其峰值与所设目标角度与极化参数一致。

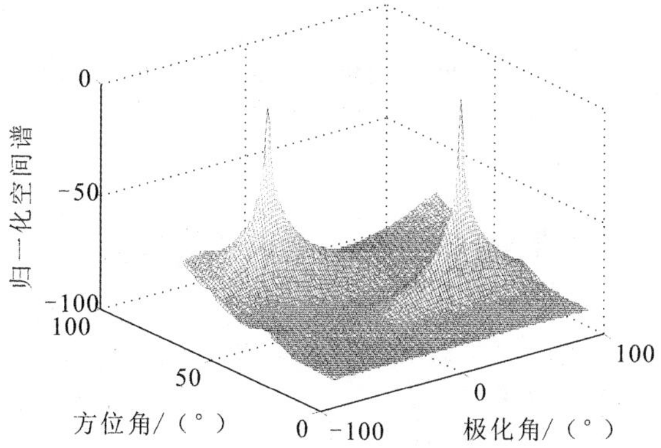

图 2.6　极化敏感阵列的 MUSIC 算法联合谱

仿真二：ESPRIT 估计结果。

信噪比 SNR 为 20 dB，快拍数为 512。图 2.7 给出了 50 次三维参数的估计结果，从图 2.7 中可以看出 ESPRIT 算法能够正确估计出角度信息和极化信息。

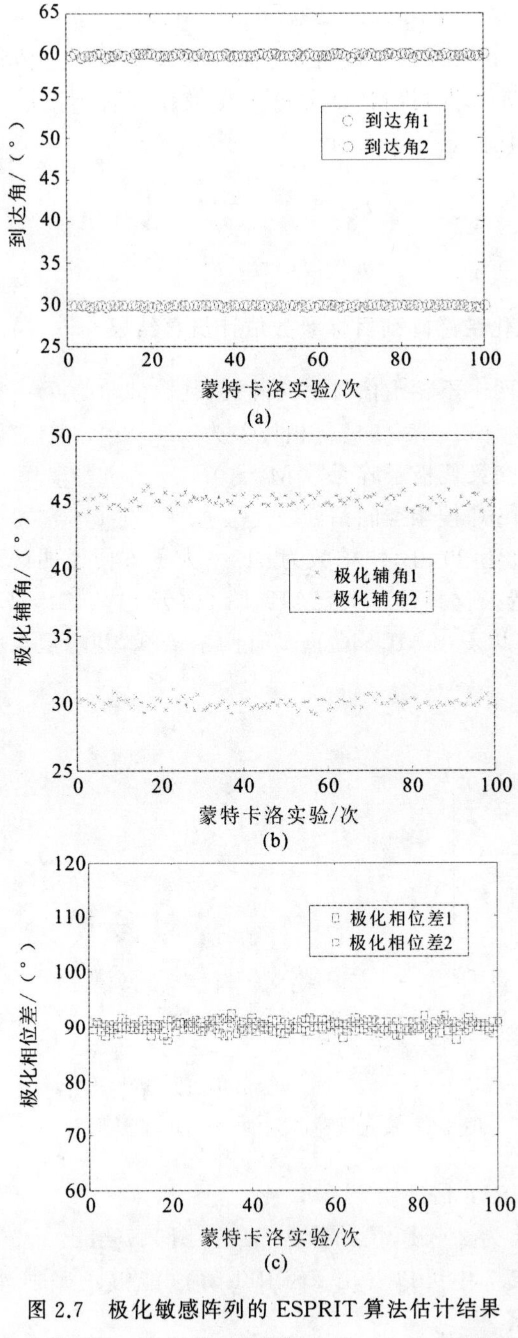

图 2.7 极化敏感阵列的 ESPRIT 算法估计结果

(a) 到达角估计结果;(b) 极化辅角估计结果;(c) 极化相位差估计结果

仿真三：ESPRIT 和 MUSIC 算法参数估计均方根误差随 SNR 的变化。

首先给出均方根误差的计算公式为

$$\mathrm{RMSE}=\sqrt{\frac{1}{\mathrm{Monte}}\sum_{m=1}^{\mathrm{Monte}}(\hat{\theta}_{m,i}-\theta_i)^2} \tag{2.30}$$

其中，Monte 为独立实验次数，也叫蒙特卡洛实验次数。$\hat{\theta}_{m,i}$ 为第 i 个目标与第 m 次的 DOA 估计值。θ_i 为第 i 个目标的 DOA 真实值。极化的参数估计 RMSE 也用类似的定义。设置快拍数为 512，信噪比在 0～30 dB 之间以步进为 5 进行变化，在每个实验数据点上都做 100 次独立实验，并利用式(2.30)求得估计的 RMSE。观察信噪比对 MUSIC 算法和 ESPRIT 算法的参数估计均方根误差的影响。仿真结果如图 2.8 所示。

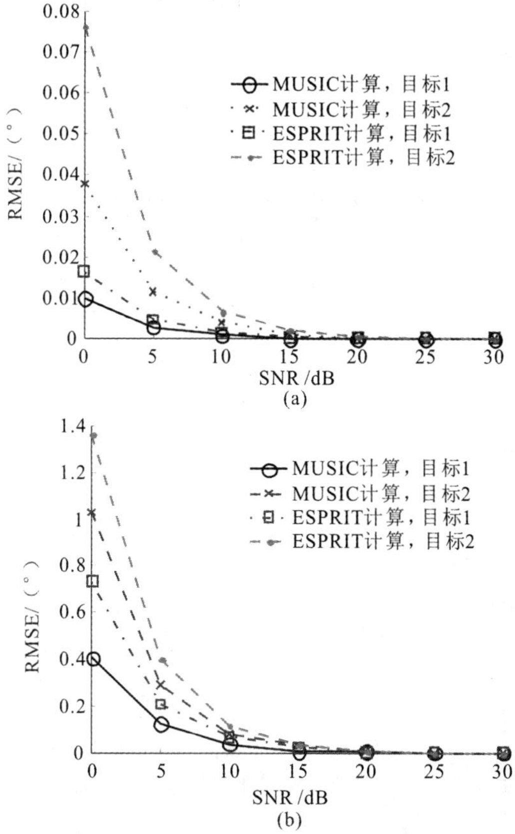

图 2.8　参数估计均方根误差与 SNR 的关系

(a)DOA 估计结果；(b)极化相位差估计结果

由图 2.8 可以得到，DOA 参数估计的 RMSE 随着信噪比的增大变得越来越小，这说明极化 MUSIC 算法以及 ESPRIT 算法的精确度不断提高。

仿真四：ESPRIT 和 MUSIC 算法参数估计均方根误差随快拍数的变化。

信噪比设置为 20 dB，快拍数从 50 逐渐增加到 1 000，每个步长下进行 100 次蒙特卡洛实验，探究步长对 DOA 参数估计均方根误差的影响，仿真结果如图 2.9 所示。

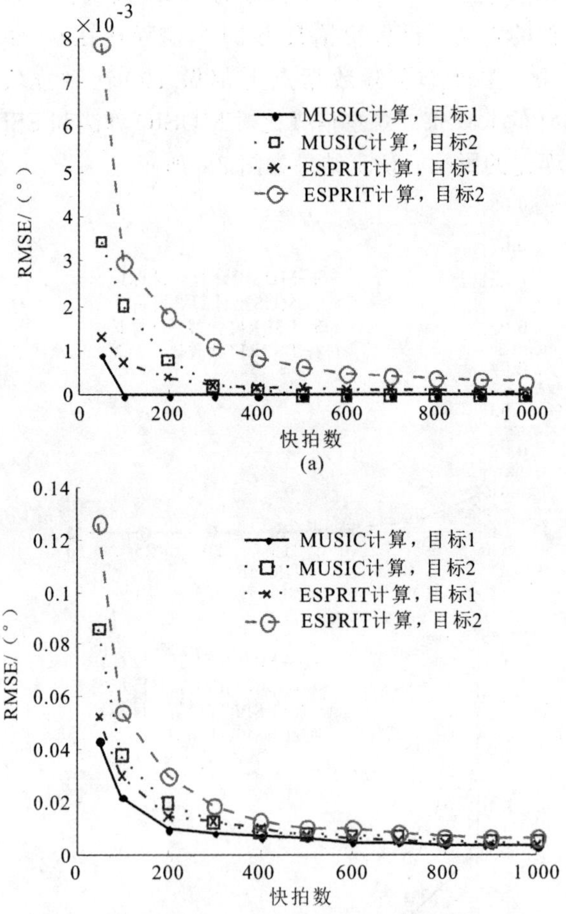

图 2.9 DOA 参数估计均方根误差与快拍数的关系

(a)DOA 估计结果；(b)极化相位差估计结果

由图 2.9 可知，随着快拍数的增大，参数估计的均方根误差逐渐减小，此结果也比较符合预期。

§2.2　波形分集技术的 MIMO 雷达参数估计基础

§2.2.1　IMO 雷达的基本知识[120][175]

通过第 1 章绪论的介绍，我们知道 MIMO 雷达是一种发射能量更加分散、虚拟孔径大的新体制雷达。与相控阵雷达不一样，MIMO 雷达采用波形不同的发射信号在空中形成所需要的波束形状，最典型的就是发射正交波形，在全空域形成等功率的波束来照射目标，然后利用阵列接收照射目标的回波信号，从而形成较大的虚拟孔径。MIMO 系统最早开发应用在无线通信领域中，在无线通信链路的两端均采用多天线发射与接收，能够做到空间分集，可成倍地提高通信系统的容量与可靠性。同样地，利用空间信息来解决传统雷达目标 RCS 起伏大的问题，Fishler 教授等人在 MIMO 通信系统的基础上提出了 MIMO 雷达的概念，后续很多学者进一步研究发现，它在目标检测、跟踪、参数估计、识别成像等方面具有相控阵雷达所不具备的优势。

绪论中已经说明 MIMO 雷达可分为统计式 MIMO 雷达和集中式 MIMO 雷达。首先说明统计式 MIMO 雷达，它的发射天线间隔很远，有数千米、数十千米甚至于数百千米，这使得每个发射天线对目标的照射均处于不同的角度，进而使发射信号经目标反射后的回波信号与统计信息不一致，甚至是不相关的，从而能比较有效地对抗目标 RCS 闪烁的情况，使得雷达的检测、跟踪、参数估计与成像等性能显著提升。统计式 MIMO 雷达对抗 RCS 闪烁的情况如图 2.10 所示。

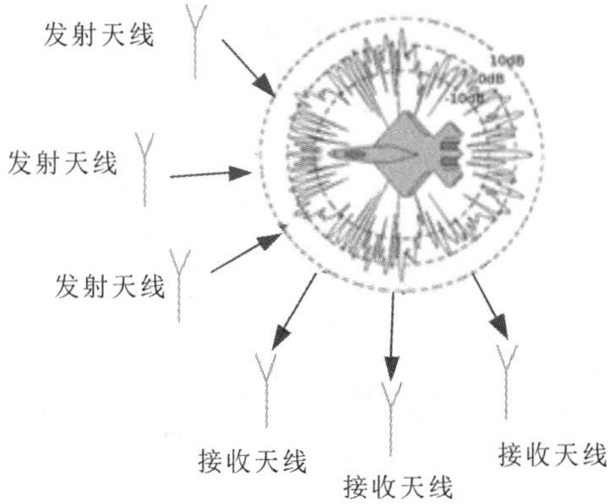

图 2.10　统计式 MIMO 雷达对抗 RCS 闪烁的情况

第二种是集中式 MIMO 雷达，它的天线间距较小，与传统相控阵雷达相似，匹配滤波后可得更大的有效虚拟孔径，则参数估计的分辨率和估计精度更高，本书侧重于参数估计，故主要研究这种体制。另外 MIMO 雷达系统的波束由于波形的独特性，其方向图的设计更加灵活。目前国内有一款典型的集中式 MIMO 雷达，它由中国电子科技集团公司第三十八研究所和西安电子科技大学联合开发，如图 2.11 所示。

图 2.11　米波稀布阵综合脉冲孔径雷达实验系统——多载频集中式 MIMO 雷达

MIMO 雷达的阵列结构有多种方式，有收发共置的，给出两种收发扩展虚拟孔径的图，如图 2.12 所示。

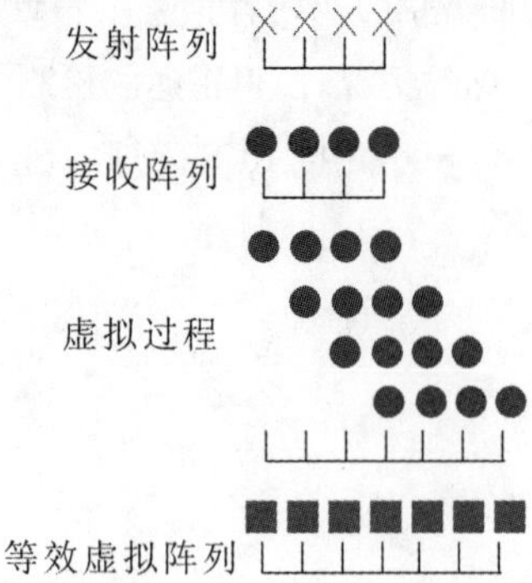

(a)

图 2.12　集中式 MIMO 雷达阵列虚拟孔径扩展

(a)密集布阵等效孔径情况；

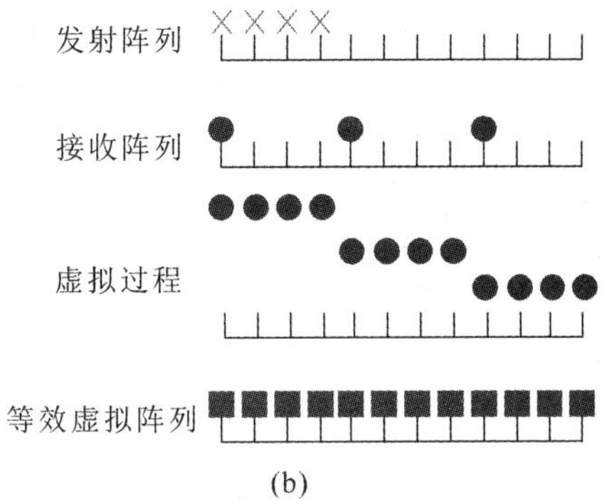

续图 2.12　集中式 MIMO 雷达阵列虚拟孔径扩展
(b)稀疏布阵等效孔径情况

§ 2.2.2　MIMO 雷达信号模型

从绪论中可知 MIMO 雷达是很热门的一个研究课题。它先发射正交信号，然后通过接收端的匹配滤波，能够产生多于接收阵元的虚拟孔径，从而提高角度估计精度。很多学者对此进行了研究，取得到了较好的研究结果。对于正交发射波形的 MIMO 雷达，其能量在全空域范围是均匀分布的，这意味着其形成的发射波束是全空域等功率形状，与传统相控阵雷达的针状波束有着天壤之别。

假设 MIMO 阵列雷达的发射和接收阵元分别为 M 和 N。MIMO 雷达符合远场条件并有 K 个目标在同一个距离单元里，其 DOA 等待估计，如图 2.13 所示。

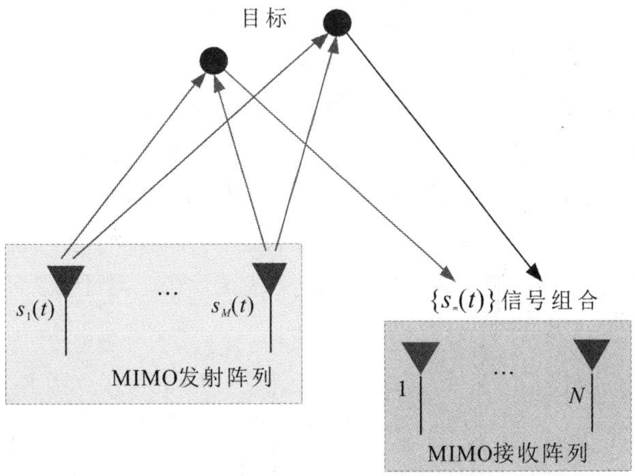

图 2.13　相干 MIMO 雷达示意图[120]

发射信号及其匹配过程如图2.14所示,具体推导如下。

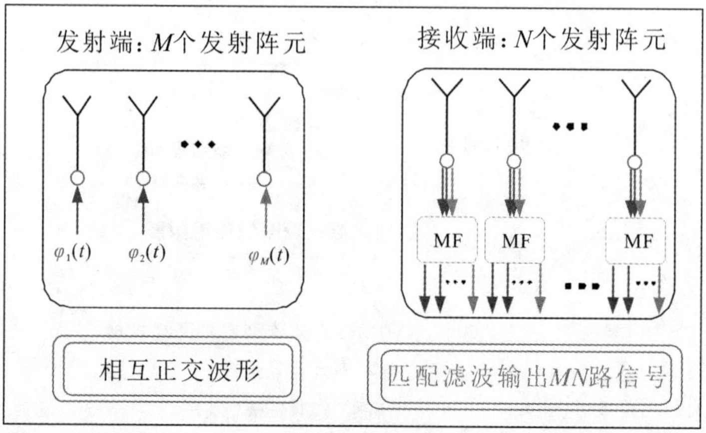

图2.14 MIMO雷达发射/接收正交波形示意图

其接收信号模型为

$$\boldsymbol{x}(t,\tau)=\boldsymbol{A}_r(\theta)\mathrm{diag}[\boldsymbol{b}(\tau)]\boldsymbol{A}_t^{\mathrm{T}}(\theta)\boldsymbol{\varphi}(t)+\boldsymbol{w}(t,\tau)\in\mathbb{C}^{N\times1} \tag{2.31}$$

其中,$\boldsymbol{A}_t(\theta)=[\boldsymbol{\alpha}_t(\theta_1)\ \cdots\ \boldsymbol{\alpha}_t(\theta_K)]\in\mathbb{C}^{M\times K}$,类似于传统相控阵雷达,可将其定义为发射导向矩阵,$\boldsymbol{A}_r(\theta)=[\boldsymbol{\alpha}_r(\theta_1)\ \cdots\ \boldsymbol{\alpha}_r(\theta_K)]\in\mathbb{C}^{N\times K}$,同样的可将其定义为接收导向矩阵,其中每一列的值分别为

$$\boldsymbol{\alpha}_t(\theta_k)=\left[\exp(-\mathrm{j}\frac{2\pi}{\lambda}d_{t,1}\sin\theta_k)\ \cdots\ \exp(-\mathrm{j}\frac{2\pi}{\lambda}d_{t,M}\sin\theta_k)\right]^{\mathrm{T}} \tag{2.32}$$

$$\boldsymbol{\alpha}_r(\theta_k)=\left[\exp(-\mathrm{j}\frac{2\pi}{\lambda}d_{r,1}\sin\theta_k)\ \cdots\ \exp(-\mathrm{j}\frac{2\pi}{\lambda}d_{r,N}\sin\theta_k)\right]^{\mathrm{T}} \tag{2.33}$$

$d_{t,k}$和$d_{r,k}$两个参数分别是第k个发射阵元和接收阵元的空间位置。$\boldsymbol{b}(\tau)$为回波起伏模型,设为Swelling2,其值$\boldsymbol{b}(\tau)=[\gamma_1(\tau)\ \cdots\ \gamma_K(\tau)]^{\mathrm{T}}\in\mathbb{C}^{K\times1}$,其中$\gamma_k(\tau)=\alpha_k\exp(\mathrm{j}2\pi f_k\tau)$。这里需要注意的是,当发射系数$\alpha_k$为不起伏模型,即Swelling5型,且多目标多普勒频率相同时,该等效信号为相干信号。本章假设目标信号不相干。$\boldsymbol{w}(t,\tau)$设为复循环高斯白噪声,其均值为0,方差为σ_n^2。发射M个发射信号:$\boldsymbol{\varphi}=[\varphi_1(t)\ \cdots\ \varphi_M(t)]^{\mathrm{T}}$。对于发射信号其协方差矩阵可用如下形式表示:

$$\boldsymbol{R}_s=\int_0^{T_p}\boldsymbol{\varphi}(t)\boldsymbol{\varphi}^{\mathrm{H}}(t)\mathrm{d}t=\begin{bmatrix}1 & \beta_{12} & \cdots & \beta_{1M}\\ \beta_{21} & 1 & \cdots & \beta_{ij}\\ \vdots & \vdots & & \vdots\\ \beta_{M1} & \beta_{M2} & \cdots & 1\end{bmatrix} \tag{2.34}$$

其中，T_p 为脉冲持续时间。利用下式的匹配滤波对接收信号进行处理：

$$\boldsymbol{X}(\tau)=\int_0^{T_p}\boldsymbol{x}(t,\tau)\boldsymbol{\varphi}(t)^{\mathrm{H}}\mathrm{d}t=\boldsymbol{A}_r(\theta)\mathrm{diag}[\boldsymbol{b}(\tau)]\boldsymbol{A}_t^{\mathrm{T}}(\theta)\boldsymbol{R}_s+\boldsymbol{N}(\tau)\in\mathbb{C}^{N\times M} \tag{2.35}$$

其中，噪声 $\boldsymbol{N}(\tau)=\int_0^{T_p}\boldsymbol{w}(t,\tau)\boldsymbol{\varphi}(t)^{\mathrm{H}}\mathrm{d}t$。当 $\beta_{ij}=0,i\neq j$，即 M 个发射信号完全正交时，$\boldsymbol{R}_s=\boldsymbol{I}_M$，对匹配滤波后接收信号进行矢量化：

$$\left.\begin{aligned}&\boldsymbol{y}(\tau)=\mathrm{vec}\{\boldsymbol{X}(\tau)\}=\boldsymbol{A}(\theta)\boldsymbol{b}(\tau)+\boldsymbol{n}(\tau)\in\mathbb{C}^{MN\times 1}\\&\boldsymbol{A}(\theta)=\boldsymbol{A}_t(\theta)\oplus\boldsymbol{A}_r(\theta)=[\boldsymbol{\alpha}(\theta_1)\ \boldsymbol{\alpha}(\theta_2)\ \cdots\ \boldsymbol{\alpha}(\theta_K)]\\&\boldsymbol{n}(\tau)=\int_0^{T_p}\boldsymbol{\varphi}(t)^{*}\otimes\boldsymbol{w}(t,\tau)\mathrm{d}t\end{aligned}\right\} \tag{2.36}$$

式(2.36)可形成孔径扩展的虚拟阵元。且已证明噪声 $\boldsymbol{n}(\tau)$ 依然为白噪声。

§2.2.3　MIMO 雷达经典角度估计算法

1.MUSIC 算法[85][86]

与上述经典算法相同，这里以 MUSIC 和 ESPRIT 超分辨算法为例来说明其在参数估计方面的优势。

先利用最大似然估计方法：$\hat{\boldsymbol{R}}=1/L\sum_{l=1}^{L}\boldsymbol{y}(t_l)\boldsymbol{y}(t_l)^{\mathrm{H}}$ 计算得到式(2.36)接收数据的协方差矩阵，其中，L 是快拍数。通过对极化敏感阵列的分析，我们知道对矩阵 $\hat{\boldsymbol{R}}$ 进行特征值分解可得 $\hat{\boldsymbol{R}}=\boldsymbol{E}_S\boldsymbol{\Sigma}_S\boldsymbol{E}_S^{\mathrm{H}}+\sigma_n^2\boldsymbol{E}_N\boldsymbol{E}_N^{\mathrm{H}}$，其中 $\boldsymbol{E}_S$ 为信号子空间，$\boldsymbol{E}_N$ 为噪声子空间。信号子空间 $\boldsymbol{E}_S$ 与导向矩阵 $\boldsymbol{A}$ 具有相同空间，即 $\boldsymbol{E}_S=\boldsymbol{AT}$，$\boldsymbol{T}$ 定义为唯一且是非奇异的矩阵。根据导向矢量与噪声子空间正交的原理，二维 MUSIC 角度搜索算法可表示为

$$P_{\mathrm{music}}=\frac{1}{[\boldsymbol{\alpha}_t(\theta)\otimes\boldsymbol{\alpha}_r(\theta)]^{\mathrm{H}}\boldsymbol{E}_N\boldsymbol{E}_N^{\mathrm{H}}[\boldsymbol{\alpha}_t(\theta)\otimes\boldsymbol{\alpha}_r(\theta)]} \tag{2.37}$$

通过式(2.37)的角度搜索谱值，K 个目标对应的 K 个峰值，其位置即为目标的 DOA 估计值。

2.ESPRIT 算法[85][86]

采用 ESPRIT 算法是为了不对阵列进行降维处理，对于 MIMO 阵列雷达的角度估计，这里采取双基地 MIMO 阵列雷达的形式，接收信号模型与上述

单基地 MIMO 雷达唯一的区别是分发射角和接收角不一致，即

$$\left.\begin{array}{l}\theta=\varphi\text{，单基地 MIMO 阵列雷达}\\ \theta\neq\varphi\text{，双基地 MIMO 阵列雷达}\end{array}\right\}\tag{2.38}$$

为了方便采用旋转子空间不变算法，假设发射阵列和接收阵列均为均匀线阵。首先，取 MIMO 阵列雷达的发射阵列前 $M-1$ 和发射阵列的后 $M-1$ 阵元构成两个完全相同的空域旋转不变子阵。那么对于第 k 个目标，该 MIMO 均匀线阵阵列雷达具有如下空域旋转不变方程：

$$\boldsymbol{J}_{\text{MIMO},2}\boldsymbol{\alpha}(\theta_k,\phi_k)=\mathrm{e}^{-\mathrm{j}2\pi d\sin\theta_k/\lambda}\boldsymbol{J}_{\text{MIMO},1}\boldsymbol{\alpha}(\theta_k,\phi_k)\tag{2.39}$$

其中，选择矩阵的值为 $\boldsymbol{J}_{\text{MIMO},1}=\boldsymbol{I}_N\otimes[\boldsymbol{I}_{(M-1)}\ \mathbf{O}_{(M-1)\times1}]$，$\boldsymbol{J}_{\text{MIMO},2}=\boldsymbol{I}_N\otimes[\mathbf{O}_{(M-1)\times1}\ \boldsymbol{I}_{(M-1)}]$。对于 K 个目标的情况，可将式(2.39)表达成矩阵相乘的形式，即

$$\boldsymbol{J}_{\text{MIMO},2}\mathbf{A}=\boldsymbol{J}_{\text{MIMO},1}\mathbf{A}\boldsymbol{\Phi}_{\text{MIMO},\theta}\tag{2.40}$$

$$\boldsymbol{\Phi}_{\text{MIMO},\theta}=\mathrm{diag}[\mathrm{e}^{-\mathrm{j}2\pi d\sin\theta_1/\lambda}\ \mathrm{e}^{-\mathrm{j}2\pi d\sin\theta_2/\lambda}\ \cdots\ \mathrm{e}^{-\mathrm{j}2\pi d\sin\theta_K/\lambda}]\tag{2.41}$$

其中，矩阵 $\boldsymbol{\Phi}_{\text{MIMO},\theta}$ 只含有未知参数 DOD 角度信息。根据信号子空间和导向矩阵的关系，式(2.40)可转化为

$$\boldsymbol{J}_{\text{MIMO},2}\boldsymbol{E}_S=\boldsymbol{J}_{\text{MIMO},1}\boldsymbol{E}_S\boldsymbol{\Psi}_{\text{MIMO},\theta}\tag{2.42}$$

其中，$\boldsymbol{\Psi}_{\text{MIMO},\theta}=\boldsymbol{T}^{-1}\boldsymbol{\Phi}_{\text{MIMO},\theta}\boldsymbol{T}$。

取 MIMO 阵列雷达的接收阵列前 $N-1$ 和接收阵列的后 $N-1$ 阵元，构成两个完全相同的空域旋转不变子阵。对于第 k 个目标，该 MIMO 均匀线阵阵列雷达具有如下空域旋转不变方程：

$$\boldsymbol{J}_{\text{MIMO},4}\boldsymbol{\alpha}(\theta_k,\phi_k)=\mathrm{e}^{-\mathrm{j}2\pi d\sin\phi_k/\lambda}\boldsymbol{J}_{\text{MIMO},3}\boldsymbol{\alpha}(\theta_k,\phi_k)\tag{2.43}$$

其中，选择矩阵的值为 $\boldsymbol{J}_{\text{MIMO},4}=[\boldsymbol{I}_{M(N-1)}\ \mathbf{O}_{M(N-1)\times1}]$，$\boldsymbol{J}_{\text{MIMO},3}=[\mathbf{O}_{M(N-1)\times1}\ \boldsymbol{I}_{M(N-1)}]$。考虑所有 K 个目标，可将式(2.43)表达成矩阵相乘的形式，即

$$\boldsymbol{J}_{\text{MIMO},4}\mathbf{A}=\boldsymbol{J}_{\text{MIMO},3}\mathbf{A}\boldsymbol{\Phi}_{\text{MIMO},\phi}\tag{2.44}$$

$$\boldsymbol{\Phi}_{\text{MIMO},\phi}=\mathrm{diag}[\mathrm{e}^{-\mathrm{j}2\pi d\sin\phi_1/\lambda}\ \mathrm{e}^{-\mathrm{j}2\pi d\sin\phi_2/\lambda}\ \cdots\ \mathrm{e}^{-\mathrm{j}2\pi d\sin\phi_K/\lambda}]\tag{2.45}$$

其中，$\boldsymbol{\Phi}_{\text{MIMO},\phi}$ 只含有未知参数 DOA 角度信息。根据信号子空间和导向矩阵的关系，可将式(2.44)转化为如下形式：

$$\boldsymbol{J}_{\text{MIMO},4}\boldsymbol{E}_S=\boldsymbol{J}_{\text{MIMO},3}\boldsymbol{E}_S\boldsymbol{\Psi}_{\theta}\tag{2.46}$$

其中，$\boldsymbol{\Psi}_{\text{MIMO},\phi}=\boldsymbol{T}^{-1}\boldsymbol{\Phi}_{\text{MIMO},\phi}\boldsymbol{T}$。

存在噪声的场景下，方程式(2.42)和式(2.46)可利用总体最小二乘算法求解得到矩阵 $\boldsymbol{\Psi}_{\text{MIMO},\theta}$，$\boldsymbol{\Psi}_{\text{MIMO},\phi}$。对 $\boldsymbol{\Phi}_{\text{MIMO},\theta}$ 和 $\boldsymbol{\Phi}_{\text{MIMO},\phi}$ 进行特征值分解，得到

其对角元数 $\{[\boldsymbol{\Phi}_{\mathrm{MIMO},\theta}]_{kk}, k=1,\cdots,K\}$ 和 $\{[\boldsymbol{\Phi}_{\mathrm{MIMO},\phi}]_{kk}, k=1,\cdots,K\}$。则 DOD 和 DOA 的估计结果为

$$\hat{\theta}=\arcsin\frac{-\lambda\angle([\boldsymbol{\Phi}_{\mathrm{MIMO},\theta}]_{kk})}{2\pi d},\quad k=1,\cdots,K \tag{2.47}$$

$$\hat{\phi}=\arcsin\frac{-\lambda\angle([\boldsymbol{\Phi}_{\mathrm{MIMO},\phi}]_{kk})}{2\pi d},\quad k=1,\cdots,K \tag{2.48}$$

§ 2.2.4　MIMO 雷达目标角度估计仿真结果

仿真一：MUSIC 空间谱。

首先考察单基地 MIMO 雷达的 MUSIC 仿真结果，发射和接收阵元数 $M=4$，$K=3$ 个独立目标存在于同一距离单元，目标 DOA 为：$\theta=\{10°,30°,50°\}$，快拍数 $L=100$，信噪比 SNR＝20 dB。阵列为半波长布阵的均匀线阵。图 2.15 给出了单基地 MIMO 雷达的 MUSIC 空间归一化谱图。从图 2.15 中可以看出，通过 MUSIC 算法，能够正确得到目标的三个谱峰。

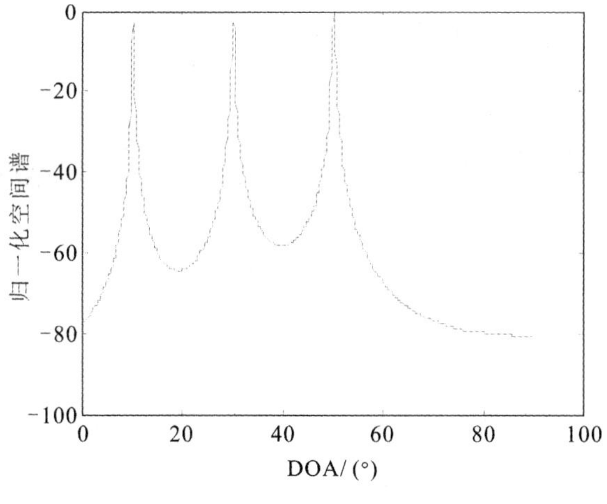

图 2.15　MIMO 雷达 MUSIC 算法空间谱

仿真二：MUSIC 算法的 RMSE 随 SNR 的变化。

仿真中，SNR 从 0 变化到 30 dB，其余仿真条件与前述相同。均方根误差(RMSE)定义为：$\mathrm{RMSE}=\sqrt{\frac{1}{\mathrm{Monte}}\sum_{k=1}^{\mathrm{Monte}}E[(\hat{\theta}_k-\theta_k)^2]}$，其中 $\hat{\theta}$ 为目标 DOA 估计值，θ 为目标 DOA 真实值。设置完成 1 000 次蒙特卡洛实验。图 2.16 给

出 MIMO 雷达的 MUSIC 算法随 RMSE 的变化曲线。从图中可以看出估计精度结果随着 SNR 的增加，其 RMSE 值逐渐减少，这表明估计性能越来越好。

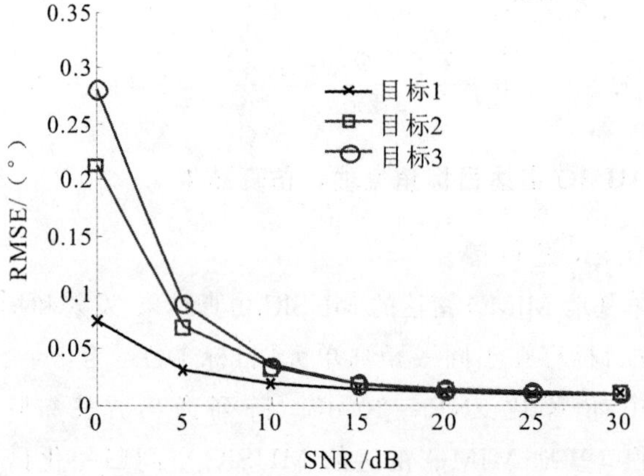

图 2.16　MIMO 雷达 MUSIC 算法随 RMSE 的变化曲线

仿真三：MUSIC 算法的 RMSE 随快拍数的变化。

本仿真中，除了快拍数从 10 变化到 200，其余仿真条件与仿真一相同。每个数据点完成 1 000 次独立实验。图 2.17 给出 MIMO 雷达 MUSIC 算法的 RMSE 估计结果。从图 2.17 中可以看出快拍数越多，RMSE 值越小。即估计性能随着快拍数变多而变好，与预期结果吻合。

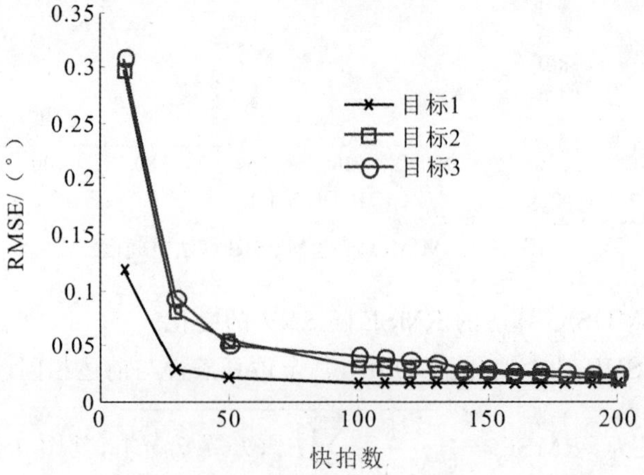

图 2.17　MIMO 雷达 MUSIC 算法的 RMSE 估计结果

仿真四：ESPRIT 算法的估计结果。

假设双基地 MIMO 阵列雷达的发射阵元数和接收阵元数分别为 $M=4$ 和 $N=4$。在同一距离单元中存在 3 个相互独立的目标，目标 DOD 和 DOA 分别为：$\theta=\{10°,20°,30°\}$，$\phi=\{20°,30°,45°\}$。信噪比和快拍数分别设置为 SNR=0 dB。$L=100$。进行 50 次独立实验，将 DOD 和 DOA 估计结果和配对结果如图 2.18 所示。从图 2.18 中可看出估计结果和角度预设值是吻合的，且目标之间的配对也是正确的。

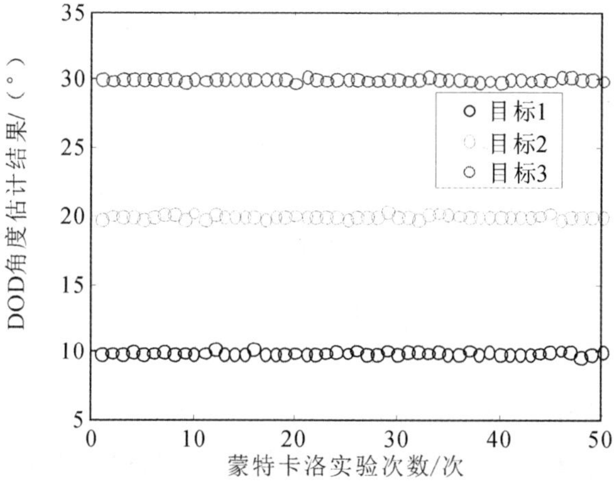

(a)

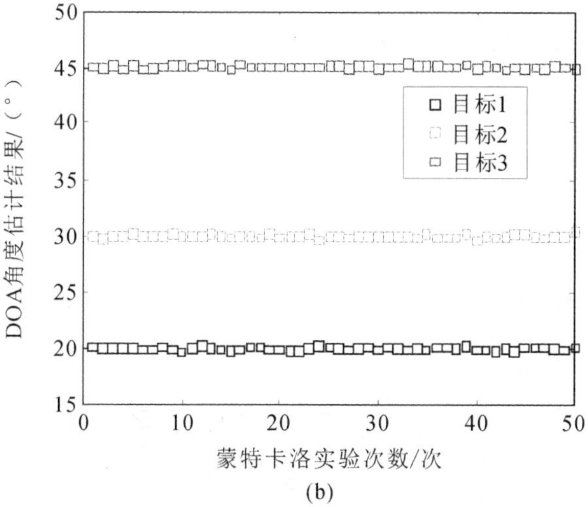

(b)

图 2.18　50 次蒙特卡洛实验值

(a) DOD 角度估计结果；(b) DOA 角度估计结果；

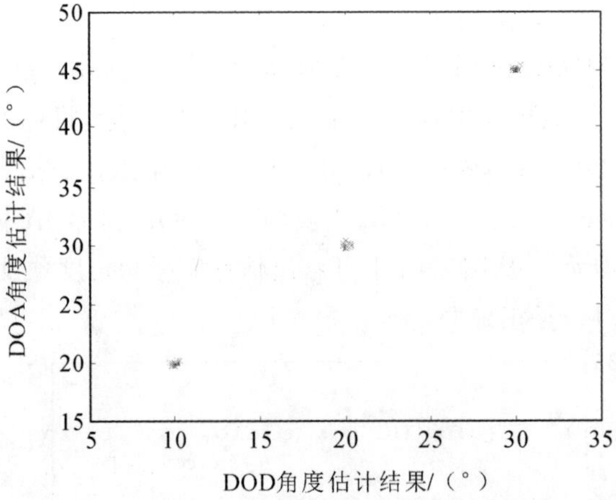

(c)

续图 2.18　50 次蒙特卡洛实验值图

(c) 配对结果

仿真五：ESPRIT 算法的 RMSE 随 SNR 的变化。

本仿真中，SNR 从 0 变化到 30 dB，其余仿真条件与上述相同。目标 DOD、DOA 和快拍数与仿真一相同，完成 1 000 次蒙特卡洛实验。目标 DOD 和 DOA 估计的 RMSE 定义与上一节类似，这里不再赘述。图 2.19 给出了目标 DOD 和 DOA 估计 RMSE 与 SNR 的关系曲线图。图 2.19 中可看出，SNR 越大，RMSE 值越小。这说明估计性能随着 SNR 的增大而提高，这是符合预期的。

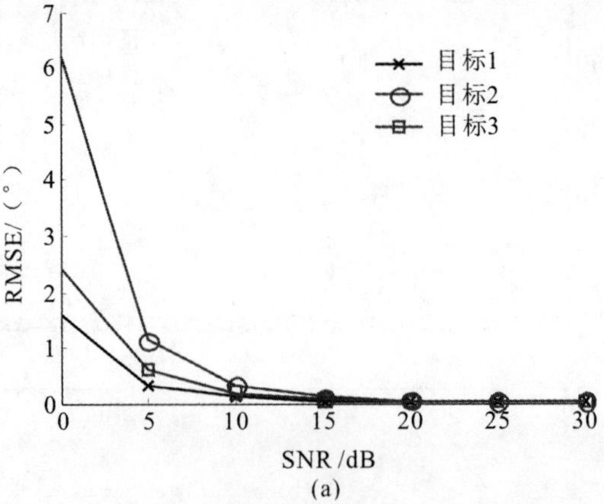

(a)

图 2.19　ESPRIT 算法的 RMSE 随 SNR 的变化

(a) DOD 的估计结果；

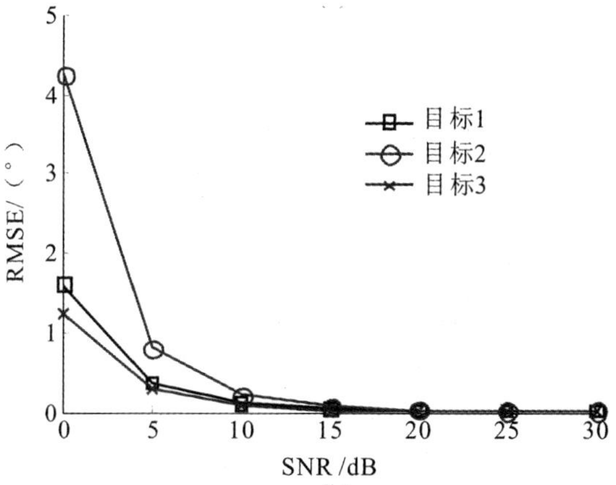

(b)

续图 2.19　ESPRIT 算法的 RMSE 随 SNR 的变化

(b) DOA 的估计结果

仿真六:ESPRIT 算法的 RMSE 随快拍数的变化。

在本仿真中,令快拍数变化,其余仿真条件设置为与仿真四相同。每个数据点完成 1 000 次独立实验。图 2.20 给出了目标 DOD 和 DOA 估计 RMSE 与快拍数关系曲线。从图 2.20 中可看出,快拍数越大,RMSE 值越小。这说明估计性能随着快拍数的增大而提高,这也是符合预期的。

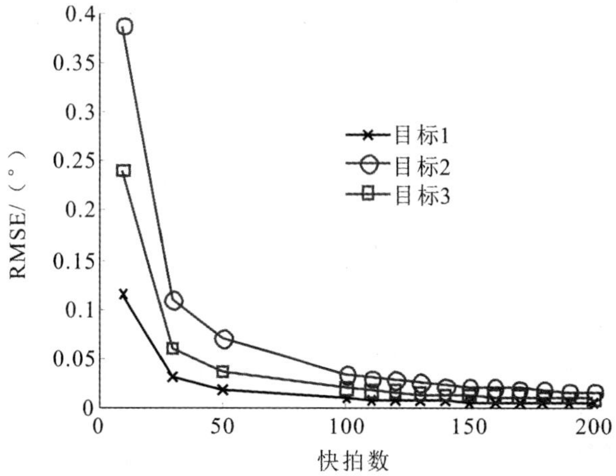

(a)

图 2.20　ESPRIT 算法的 RMSE 随快拍数的变化

(a) DOD 的估计结果;

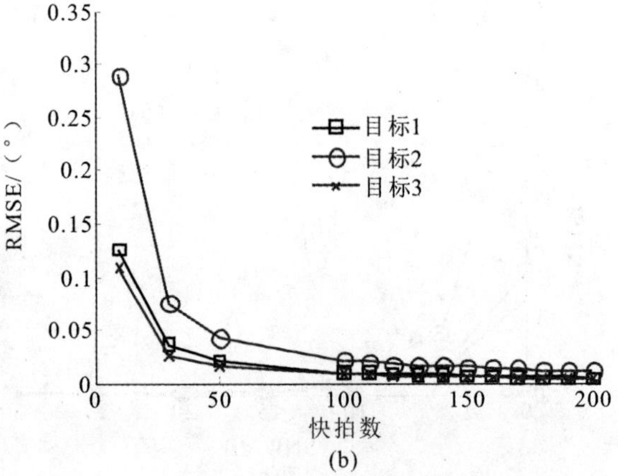

(b)

续图 2.20　ESPRIT 算法的 RMSE 随快拍数的变化
(b) DOA 的估计结果

第 3 章　基于极化分集的分离式大尺寸极化敏感阵列参数估计算法研究

§ 3.1　引　　言

极化敏感阵列能够获取极化分集，从而提高目标的参数估计精度。故基于极化敏感阵列的角度及极化参数的估计问题得到了广泛的研究。具体成果可归结为如下几个阶段。

第一阶段：共点式电磁矢量传感器条件下的各种估计算法，包括矢量叉积算法和各种经典的超分辨算法，如 ESPRIT、MUSIC 等。

第二阶段：分离式电磁矢量传感条件下的各种估计算法，包括矢量叉积算法和各种经典的超分辨算法，以及各种非理想条件下的角度估计算法。

第三阶段：共点式和分离式矢量传感器条件下，基于稀疏表征的各种角度估计算法，如 OMP、BOMP 等。

上述研究均基于短电偶极子和小磁环的条件下，但是其辐射效率太低。基于此，在 Wong 博士的引领之下，目前进入第四个阶段，即研究“长”电偶极子和“大”磁环组成的阵列的目标参数估计。Wong 博士已经给出了“长”电偶极子和“大”磁环的阵元响应，并给出了一种简单的参数估计算法[45-46]。但是需要预先知道极化信息后，才能够估计目标的二维角度，这在雷达的应用领域是不可以接受的。

本章在分离式电磁矢量传感器的基础下，给出一种基于 ESPRIT 的算法，无需预先知道极化信息即可估计出目标的二维角度估计，也同时能够得到其极化参数估计。本章的主要亮点有两个：一是将分离式矢量传感器阵列与“长”电偶极子和“大”磁环相结合，构成一个在实际工程更加实用的阵列，此阵列的好处在于辐射效率高、阵元间互耦低。二是在估计二维角度时无需预先知道极化角，这在雷达应用中是极有好处的。

§3.2 分离式大电偶极子的接收信号模型

阵列结构如图 3.1 所示。该阵列由八个电偶极子组成(这里需要注意,可由四的整数倍构成,不局限于八个)。从图中可以看出由上、下两个子阵构成,子阵一有四个电偶极子,记为:E_x、E_y、E_z 和 Ex,其中前三个电偶极子的坐标为:$E_x=(0,y_a,0)$,$E_y=(0,y_b,0)$,$E_z=(0,y_c,0)$。可以看到当 $y_a=y_b=y_c$ 时,前三个电偶极子退化为传统的三正交共点式极化矢量天线。子阵一的第四个阵元为 Ex,其与前面 E_x 的间距为 D_y,且大于半波长。子阵二的四个阵元组成的子阵为子阵一往 x 轴平行移动,移动距离为 D_x,且大于半波长;方位角 ϕ_x 的定义和俯仰角 θ_z 定义见文献[45]。

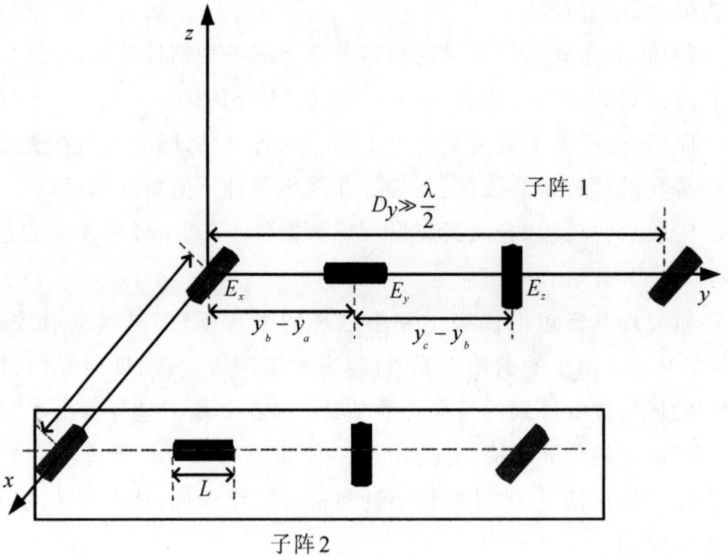

图 3.1 三正交长电偶极子阵列

下面详细推导该阵列的导向矢量。该阵列的 x 轴方向和 y 轴方向的空域导向矢量可表示为

$$\boldsymbol{\alpha}_x(\theta,\phi)=[1\ \ e^{-j\frac{2\pi}{\lambda}D_x\sin\theta_z\sin\varphi_x}]^{\mathrm{T}} \tag{3.1}$$

$$\boldsymbol{\alpha}_y(\theta_z,\phi_x)=[e^{-j\frac{2\pi}{\lambda}y_a v}\ \ e^{-j\frac{2\pi}{\lambda}y_b v}\ \ e^{-j\frac{2\pi}{\lambda}y_c v}\ \ e^{-j\frac{2\pi}{\lambda}D_y v}]^{\mathrm{T}} \tag{3.2}$$

定义 $\nu \triangleq \sin\theta_z\cos\phi_x$。除了空域导向矢量之外,还有四个长电偶极子的极化响应可以表示为[45]

$$\boldsymbol{\alpha}^{(L)}=-\begin{bmatrix}e_x\\e_y\\e_z\\e_x\end{bmatrix}\odot\begin{bmatrix}\ell_{\theta_x}^L\\\ell_{\theta_y}^L\\\ell_{\theta_z}^L\\\ell_{\theta_x}^L\end{bmatrix}\odot\begin{bmatrix}\csc\theta_x\\\csc\theta_y\\\csc\theta_z\\\csc\theta_x\end{bmatrix}\tag{3.3}$$

其中，

$$\begin{bmatrix}e_x\\e_y\\e_z\end{bmatrix}=\begin{bmatrix}\cos\theta_z\cos\varphi_x\sin\gamma\mathrm{e}^{\mathrm{j}\eta}-\sin\varphi_x\cos\gamma\\\cos\theta_z\sin\varphi_x\sin\gamma\mathrm{e}^{\mathrm{j}\eta}+\cos\varphi_x\cos\gamma\\-\sin\theta_z\sin\gamma\mathrm{e}^{\mathrm{j}\eta}\end{bmatrix}\tag{3.4}$$

$$\ell_\theta^L=-\frac{\lambda}{\pi}\frac{1}{\sin(\pi L\dfrac{1}{\lambda})}\frac{\cos\left(\pi L\dfrac{1}{\lambda}\cos\theta\right)-\cos(\pi L\dfrac{1}{\lambda})}{\sin\theta}\tag{3.5}$$

其中，θ 可以表示 θ_x，θ_y，θ_z。可以用如下的公式计算得到：

$$\sin\theta_x=\sqrt{\sin^2\theta_z\ \sin^2\phi_x+\cos^2\theta_z}\tag{3.6}$$

$$\cos\theta_x=\sin\theta_z\cos\phi_x\tag{3.7}$$

$$\sin\theta_y=\sqrt{\sin^2\theta_z\ \cos^2\phi_x+\cos^2\theta_z}\tag{3.8}$$

$$\cos\theta_y=\sin\theta_z\sin\phi_x\tag{3.9}$$

$$\sin\phi_y=\frac{\cos\theta_z}{\sqrt{\sin^2\theta_z\ \sin^2\phi_x+\cos^2\theta_z}}\tag{3.10}$$

$$\cos\phi_y=\frac{\sin\theta_z\sin\phi_x}{\sqrt{\sin^2\theta_z\ \sin^2\phi_x+\cos^2\theta_z}}\tag{3.11}$$

$$\sin\phi_z=\frac{\sin\theta_z\cos\phi_x}{\sqrt{\sin^2\theta_z\ \cos^2\phi_x+\cos^2\theta_z}}\tag{3.12}$$

$$\cos\phi_z=\frac{\cos\theta_z}{\sqrt{\sin^2\theta_z\ \cos^2\phi_x+\cos^2\theta_z}}\tag{3.13}$$

则整个阵列的导向矢量等于空域导向矢量和极化域响应的结合：

$$\boldsymbol{\alpha}(\theta_z,\phi_x,\gamma,\eta)=\boldsymbol{\alpha}_x(\theta_z,\phi_x)\otimes[\boldsymbol{\alpha}_y(\theta_z,\phi_x)\odot\boldsymbol{\alpha}^{(L)}(\theta_z,\phi_x,\gamma,\eta)]\in\mathbb{C}^{8\times1}\tag{3.14}$$

导向矢量已知，则接收数据矢量可表示为

$$\boldsymbol{x}(t)=\sum_{k=1}^{K}\boldsymbol{\alpha}(\theta_k,\phi_k,\gamma_k,\eta_k)s_k(t)+\boldsymbol{n}(t)=\boldsymbol{As}(t)+\boldsymbol{n}(t)\tag{3.15}$$

其中，K 为同一个距离单元包含的远场目标个数。根据式(3.15)知道 $\boldsymbol{A}=[\boldsymbol{\alpha}(\theta_{z,1},\phi_{x,1},\gamma_1,\eta_1)\ \cdots\ \boldsymbol{\alpha}(\theta_{z,K},\phi_{x,K},\gamma_K,\eta_K)]\in\mathbb{C}^{8\times K}$，是由 K 个导向矢

量组成的矩阵。信号矢量 $\boldsymbol{s}(t)=[s_1(t)\ \cdots\ s_K(t)]^{\mathrm{T}}\in\mathbb{C}^{K\times 1}$ 和噪声矢量 $\boldsymbol{n}(t)\in\mathbb{C}^{8\times 1}$ 设置成零均值复高斯随机分布,这是合理的,因为在雷达应用中,该信号经过脉压、MTI//MTD、恒虚警等信号处理之后,信号形式并不是已知的。

如果是大电磁环的话,则式(3.3)改成如下形式:

$$\boldsymbol{\alpha}^{(\frac{R}{\lambda})}=\begin{bmatrix}h_x\\h_y\\h_z\\h_x\end{bmatrix}\odot\begin{bmatrix}\ell_x^{(\frac{R}{\lambda})}(\theta_z,\phi_x)\\\ell_y^{(\frac{R}{\lambda})}(\theta_z,\phi_x)\\\ell_z^{(\frac{R}{\lambda})}(\theta_z,\phi_x)\\\ell_x^{(\frac{R}{\lambda})}(\theta_z,\phi_x)\end{bmatrix}\tag{3.16}$$

其中:

$$\begin{bmatrix}h_x\\h_y\\h_z\end{bmatrix}=\begin{bmatrix}-\sin\varphi_x-\cos\theta_z\cos\phi_x\\\cos\varphi_x-\cos\theta_z\sin\phi_x\\0\sin\theta_z\end{bmatrix}\begin{bmatrix}\sin\gamma\mathrm{e}^{\mathrm{j}\eta}\\\cos\gamma\end{bmatrix}\tag{3.17}$$

$$\ell_x^{(\frac{R}{\lambda})}(\theta_z,\phi_x)=\mathrm{j}2\pi R\frac{J_1\left(2\pi\dfrac{R}{\lambda}\sqrt{\sin^2\theta_z\ \sin^2\phi_x+\cos^2\theta_z}\right)}{\sqrt{\sin^2\theta_z\sin^2\phi_x+\cos^2\theta_z}}\tag{3.18}$$

$$\ell_y^{(\frac{R}{\lambda})}(\theta_z,\phi_x)=\mathrm{j}2\pi R\frac{J_1\left(2\pi\dfrac{R}{\lambda}\sqrt{\sin^2\theta_z\ \cos^2\phi_x+\cos^2\theta_z}\right)}{\sqrt{\sin^2\theta_z\ \cos^2\phi_x+\cos^2\theta_z}}\tag{3.19}$$

$$\ell_z^{(\frac{R}{\lambda})}\theta_z,\phi_x=\mathrm{j}2\pi R\frac{J_1\left(2\pi\dfrac{R}{\lambda}\sin\theta_z\right)}{\sin\theta_z}\tag{3.20}$$

§3.3 基于 ESPRIT 的虚拟拟合二维 DOA 估计算法与分析

§3.3.1 虚拟高精度模糊 DOA 计算

根据式(3.15),可计算其阵列协方差矩阵为 $\boldsymbol{R}=E\{\boldsymbol{x}(t)\boldsymbol{x}^{\mathrm{H}}(t)\}$。这是理想状态,事实上并不能够得到无限次快拍,从而求得它的期望值,可用有限个快拍数据 P,并利用最大似然得到其近似值,即 $\hat{R}=\dfrac{1}{P}\sum\limits_{p=1}^{P}\boldsymbol{x}(t_p)\boldsymbol{x}^{\mathrm{H}}(t_p)$。再根据第二章中的子空间类算法的信号子空间推导过程得到维数为 $8\times K$ 的子空间 $\boldsymbol{E}_S\in\mathbb{C}^{8\times K}$。

从整个阵列的构造过程可看出子阵二是由子阵一沿 x 轴平行移动 D_x 的距离得到的，即子阵一和子阵二具有空域旋转变性，可利用超分辨 ESPRIT 算法求得某些参数值(见图 3.1)。该旋转不变性反应在阵列导向矢量中为

$$\boldsymbol{J}_2\boldsymbol{\alpha}(\theta_z,\phi_x,\gamma,\eta)=\mathrm{e}^{-\mathrm{j}2\pi D_x u/\lambda}\boldsymbol{J}_1\boldsymbol{\alpha}(\theta_z,\phi_x,\gamma,\eta) \tag{3.21}$$

其中，$\boldsymbol{J}_2=[\boldsymbol{O}_{4\times4}\ \boldsymbol{I}_4]$和 $\boldsymbol{J}_1=[\boldsymbol{I}_4\boldsymbol{O}_{4\times4}]$被定义为选择矩阵，其功能为选择导向矢量中的前四个元素和后四个元素。如果考虑全部的目标个数，那么式(3.21)可用堆栈的思想转化为如下的矩阵关系：

$$\boldsymbol{J}_2\boldsymbol{A}=\boldsymbol{J}_1\boldsymbol{A\Phi} \tag{3.22}$$

其中，$\boldsymbol{\Phi}=\mathrm{diag}[\mathrm{e}^{-\mathrm{j}2\pi D_x u_1/\lambda}\ \mathrm{e}^{-\mathrm{j}2\pi D_x u_2/\lambda}\cdots\mathrm{e}^{-\mathrm{j}2\pi D_x u_K/\lambda}]$。无噪条件下，把 $\boldsymbol{E}_S=\boldsymbol{AT}$ 代入式(3.22)可得

$$\boldsymbol{J}_2\boldsymbol{E}_S=\boldsymbol{J}_1\boldsymbol{E}_S\boldsymbol{\Psi} \tag{3.23}$$

其中，$\boldsymbol{\Psi}=\boldsymbol{T}^{-1}\boldsymbol{\Phi T}$。

存在噪声的场景下，方程(3.23)可利用总体最小二乘算法求解得到矩阵 $\boldsymbol{\Psi}$。对矩阵 $\boldsymbol{\Psi}$ 进行特征值分解，得到其对角元素$\{[\boldsymbol{\Phi}]_{kk},k=1,\cdots,K\}$。因为子阵间的间距 D_x 大于半波长，利用式(3.24)可计算得到方向余弦 u_k^{fine} 的估计值，但是其存在周期性模糊，即

$$u_k^{\mathrm{fine}}=\frac{\angle[\boldsymbol{\Phi}]_{kk}}{2\pi D_x/\lambda},\quad k=1,\cdots,K \tag{3.24}$$

根据已有文献的结论，方向余弦 u_k 的周期模糊值可表示为

$$u_k^{\mathrm{fine},(m)}=u_k^{\mathrm{fine}}+m\lambda/D_x,\lceil(-1-u_k^{\mathrm{fine}})D_X/\lambda\rceil\leqslant m\leqslant\lfloor(1-u_k^{\mathrm{fine}})D_x/\lambda\rfloor \tag{3.25}$$

定义 m 的个数为 Q。这样利用 ESPRIT 就得到了一个方向余弦的估计，要想估计二维目标必须得到另一维的方向余弦的估计。下面来推导，首先可以通过反推得到导向矢量的估计值：

$$\boldsymbol{A}=\begin{bmatrix}\boldsymbol{J}_1\boldsymbol{A}\\ \boldsymbol{J}_2\boldsymbol{A}\end{bmatrix}=\begin{bmatrix}\boldsymbol{J}_1\boldsymbol{A}\\ \boldsymbol{J}_1\boldsymbol{A\Phi}\end{bmatrix}=\boldsymbol{E}_S\boldsymbol{T}^{-1} \tag{3.26}$$

利用关系式(3.26)，可得

$$\boldsymbol{J}_1\hat{\boldsymbol{A}}=\frac{1}{2}(\boldsymbol{J}_1\boldsymbol{E}_S\boldsymbol{T}^{-1}+\boldsymbol{J}_2\boldsymbol{E}_S\boldsymbol{T}^{-1}\boldsymbol{\Phi}^{-1})=[\hat{\boldsymbol{\alpha}}_{y,1}\cdots\hat{\boldsymbol{\alpha}}_{y,K}] \tag{3.27}$$

其中，$\hat{\boldsymbol{\alpha}}_{y,k}$是子阵一的导向矢量。紧接着，利用子阵中两个相同的阵元比值可以得到另一维方向余弦，这也是子阵如此设置的原因。

$$\nu_k^{\mathrm{fine}}=\frac{1}{2\pi D_y y/\lambda}\angle\frac{[\hat{\boldsymbol{\alpha}}_{y,k}]_4}{[\hat{\boldsymbol{\alpha}}_{y,k}]_1},\quad k=1,\cdots,K \tag{3.28}$$

因为设置的 D_y 大于半波长，利用式(3.29)可计算得到方向余弦 ν_k^{fine} 的估计值，但是其同样存在周期性模糊，即

$$\nu_k^{\text{fine},(n)}=\nu_k^{\text{fine}}+n\lambda/D_Y,\lceil(-1-\nu_k^{\text{fine}})D_y/\lambda\rceil\leqslant n\leqslant\lfloor(1-v_k^{\text{fine}})D_Y/\lambda\rfloor \tag{3.29}$$

定义 n 的个数为 R。式(3.25)和式(3.29)中 $u_k^{\text{fine},(m)}$ 和 $v_k^{\text{fine},(n)}$ 的估计值可通过简易计算得到二维 DOA$\{\hat{\theta}_k^{\text{fine},(m,n)},\hat{\varphi}_k^{\text{fine},(m,n)}\}$。该组估计值是采用阵元间的空域信息得到的二维角度估计。下面用阵元间的极化信息来排除或者找出其中一组正确的估计值。

§3.3.2 虚拟低精度模糊 DOA 计算与解模糊

将上述得到的二维估计角度代入三正交偶极子的导向矢量中，补偿掉空间相移因子和长电偶极子所带来的阵元间不一致的响应。即

首先利用 $\{\hat{\theta}_{z,k}^{\text{fine},(m,n)},\hat{\phi}_{x,k}^{\text{fine},(m,n)}\}$ 计算得到 $\sin(\hat{\theta}_{x,k}^{\text{fine},(m,n)})$，$\sin(\hat{\theta}_{y,k}^{\text{fine},(m,n)})$，$\sin(\hat{\theta}_{z,k}^{\text{fine},(m,n)})$ 的值，然后计算 $\frac{1}{\ell_{\theta_x}^L},\frac{1}{\ell_{\theta_y}^L},\frac{1}{\ell_{\theta_z}^L}$ 的估计值。将上述估计值代入三正交偶极子导向矢量，以补偿空间相移因子和长电偶极子引起的单元间不一致响应，得到三正交集中式短电偶极子的导向矢量估计值：

$$\begin{bmatrix}\hat{e}_x\\ \hat{e}_y\\ \hat{e}_z\end{bmatrix}_{k,(m,n)}=[\boldsymbol{J}_3\hat{\boldsymbol{\alpha}}_y(\theta_k,\phi_k,\gamma_k,\eta_k)]\odot\begin{bmatrix}\mathrm{e}^{-\mathrm{j}\frac{2\pi}{\lambda}(y_a\hat{v}_k^{(n)})}\\ \mathrm{e}^{-\mathrm{j}\frac{2\pi}{\lambda}(y_b\hat{v}_k^{(n)})}\\ \mathrm{e}^{-\mathrm{j}\frac{2\pi}{\lambda}(y_b\hat{v}_k^{(n)})}\end{bmatrix}\odot\begin{bmatrix}\frac{1}{\ell_{\theta_x}^L}\\ \frac{1}{\ell_{\theta_y}^L}\\ \frac{1}{\ell_{\theta_z}^L}\end{bmatrix}\odot\begin{bmatrix}\sin(\hat{\theta}_{x,k}^{\text{fine},(m,n)})\\ \sin(\hat{\theta}_{y,k}^{\text{fine},(m,n)})\\ \sin(\hat{\theta}_{z,k}^{\text{fine},(m,n)})\end{bmatrix} \tag{3.30}$$

其中，$\boldsymbol{J}_3=[\boldsymbol{I}_3\ \mathbf{O}_{3,1}]$。式(3.30)得到的估计值即为三正交同心短电偶极子的导向矢量估计值，则利用文献[40]的结果得到其对应的 DOA 粗估计值 $\{\phi_k^{\text{coarse},(m,n)},\theta_k^{\text{coarse},(m,n)}\}$，显然该组估计值中有且只有一组估计值是正确的。所以将精估计值和该组粗估计值作对比，找到最相同的一组估计值，即为目标的二维估计值，可经下式得到：

$$(m^o,n^o)=\arg\min_{m,n}\{[\theta_k^{\text{coarse},(m,n)}+\phi_k^{\text{coarse},(m,n)}-\theta_k^{\text{fine},(m,n)}-\phi_k^{\text{fine},(m,n)}]^2\} \tag{3.31}$$

则目标俯仰角和方位角的估计值为

$$\{\hat{\theta}_k=\theta_k^{\text{fine},(m^o,n^o)},\hat{\varphi}_k=\varphi_k^{\text{fine},(m^o,n^o)}\},k=1,\cdots,K \tag{3.32}$$

根据上面的分析，可以将角度估计算法总结如下：

(1)找到两个子阵之间的空域旋转不变性，利用 ESPRIT 超分辨算法得到其中一维的方向余弦周期性模糊高精度估计值。

(2)利用子阵间第一个元素和第四个元素的比值可得到另一维的方向余弦周期性模糊高精度估计值，利用得到的二维方位余弦估计值得到周期性模糊高精度二维 DOA 估计值。

(3)子阵中的前三个阵元存在空间相移因子，利用(2)中得到的二维 DOA 进行相移因子补偿，若补偿正确，则可得到三正交共点式电偶极子矢量天线，若补偿不正确，则得到另外虚拟的分离式矢量天线。利用补偿的数据得到多组二维 DOA 估计值，需要注意的是只有补偿正确的那一组数据，得到的二维 DOA 估计值才是正确且无模糊。

(4)将(2)中的二维 DOA 精估计值与(3)的二维 DOA 粗估计值进行对比，寻找到误差最小的那一组，理论上误差最小的那一组估计值代表了正确的二维 DOA 精度值和正确的二维 DOA 粗估计值。取这一步的二维 DOA 精估计值作为最终二维 DOA 估计值。

为了更好地理解算法步骤中的内容，下面画出每个步骤所对应的示意图，如图 3.2 所示。

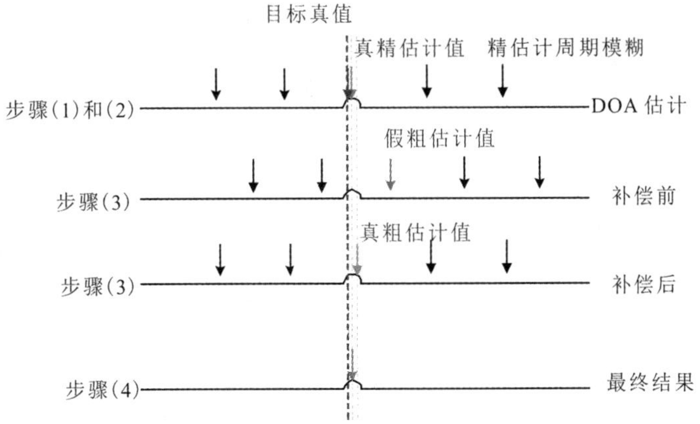

图 3.2　本章所提算法角度估计示意图

该算法的优点有：一是无需提前知道角度或者极化的先验知识。二是分离式极化敏感阵列阵元间互耦影响较小。三是分离式阵列结构可以任意。四是阵元间距远大于半波长，即有效孔径大、估计精度高。五是算法无需角度搜索，计算量小。本章所提算法缺点有两个：一是需要方位和俯仰角的估计配对操作。二是在粗估计精度不够高时，最终角度估计精度迅速下降。

§3.3.3 本章算法适用的极化敏感阵列

完整的电磁矢量传感器包含 6 个分量，即：E_x、E_y 和 E_z 三个电场分量，H_x，H_y 和 H_z 三个磁场分量，从上述算法推导过程中可以看到，估计过程中唯一受到算法限制的为式(3.30)。即当式(3.30)在相位补偿之后能够利用文献中的算法即可。其子阵包含的 4 个极化分量必须为：X_x、X_y 和 X_z 三个电场或者磁场分量，其中 X 既可以是电场 E 也可以是磁场 H，子阵的最后一个分量是为得到模糊精估计而设计的，故根据式(3.28)知道只要第四个分量与前面三个任意一个分量相同即可。图 3.3 和图 3.4 给出两个典型的子阵结构，图 3.3 为全大磁环组成子阵，图 3.4 为长电偶极子和大磁环混编阵列结构。另外，该算法可扩展至多个子阵，并不受限于两个四阵元组成的子阵，图 3.5 给出了多个子阵构成的阵列形式。

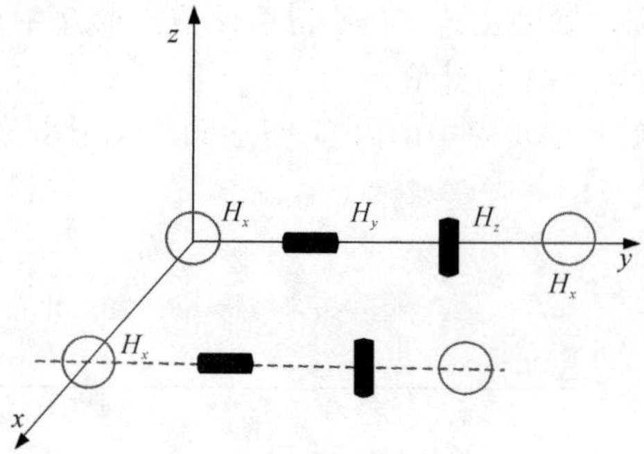

图 3.3 大磁环子阵阵列结构

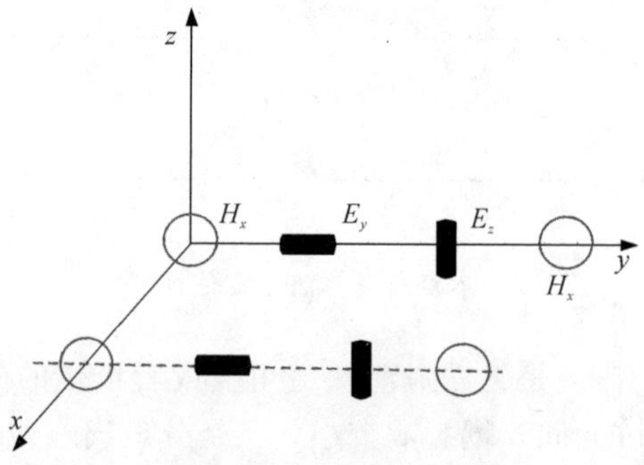

图 3.4 长电偶极子和大磁环混编阵列结构

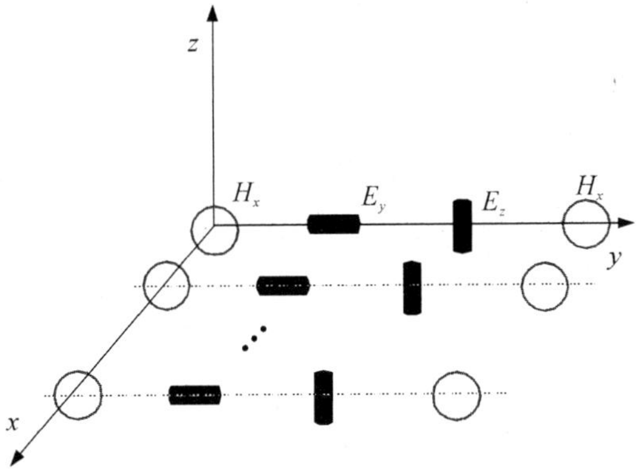

图 3.5　长电偶极子和大磁环混编的多子阵阵列结构

对于所有阵元均是大电磁环的情况，如图 3.3 所示的阵列结构，只需两步修改，第一是对 ESPRIT 算法进行改进，以估计方向余弦。第二，如果阵列是由磁环组成的，则公式(3.30)可以改为

$$\begin{bmatrix} \hat{h}_x \\ \hat{h}_y \\ \hat{h}_z \end{bmatrix}_{k,(m,n)} = \boldsymbol{J}_3 \hat{\boldsymbol{\alpha}}_y(\theta_k, \varphi_k, \gamma_k, \eta_k) \odot \begin{bmatrix} \mathrm{e}^{-\mathrm{j}\frac{2\pi}{\lambda}(y_a \hat{\nu}_k(n))} \\ \mathrm{e}^{-\mathrm{j}\frac{2\pi}{\lambda}(y_b \hat{\nu}_k(n))} \\ \mathrm{e}^{-\mathrm{j}\frac{2\pi}{\lambda}(y_b \hat{\nu}_k(n))} \end{bmatrix} \odot \begin{bmatrix} \dfrac{1}{\hat{\ell}_x\left(\frac{R}{\lambda}\right)(\theta_z, \phi_x)} \\ \dfrac{1}{\hat{\ell}_y\left(\frac{R}{\lambda}\right)(\theta_z, \phi_x)} \\ \dfrac{1}{\hat{\ell}_z\left(\frac{R}{\lambda}\right)(\theta_z, \phi_x)} \end{bmatrix} \tag{3.33}$$

§3.3.4　计算复杂度

该方法的计算复杂度主要由四个步骤决定。第一步是利用 ESPRIT 算法计算轴方向余弦。第二步是计算另一个垂直轴的方向余弦。第三步是计算分离式和长电偶极子导向矢量的补偿。第四步是对粗估计和精估计进行拟合。

第一步可以分为三个小步骤：第一小步为计算协方差矩阵需要 $O\{M^2 P\}$ 复数次乘法运算；第二小步特征值分解需要 $O\{(M)^3\}$；第三小步需要 $O\left\{\left(\frac{M}{2}\right)^2 K+K^3\right\}$。其中：第二步需要 $O\left\{\frac{M}{2}K^2+K^3\right\}$；第三步需要 $O\{9K\}$；

第四步则需要 $O\{KQR\}$。QR 分别在式(3.25)和(3.29)中定义。

§ 3.3.5 CRB

从信号模型中可以看出带估计参数为$\{\theta_{z,k},\phi_{x,k},\gamma_k,\eta_k\}$。定义 $\boldsymbol{\theta}=[\theta_{z,1}\ \theta_{z,2}\cdots\theta_{z,K}]$,$\boldsymbol{\phi}=[\phi_{x,1}\ \phi_{x,2}\cdots\phi_{x,K}]$,$\boldsymbol{\gamma}=[\gamma_1\ \gamma_2\cdots\gamma_K]$,$\boldsymbol{\eta}=[\eta_1\ \eta_2\cdots,\eta_K]$。然后其 FIM 可以表示为

$$\boldsymbol{J}=\begin{bmatrix}\boldsymbol{J}_{\theta\theta} & \boldsymbol{J}_{\theta\phi} & \boldsymbol{J}_{\theta\gamma} & \boldsymbol{J}_{\theta\eta}\\ \boldsymbol{J}_{\phi\theta} & \boldsymbol{J}_{\phi\phi} & \boldsymbol{J}_{\phi\gamma} & \boldsymbol{J}_{\phi\eta}\\ \boldsymbol{J}_{\gamma\theta} & \boldsymbol{J}_{\gamma\phi} & \boldsymbol{J}_{\gamma\gamma} & \boldsymbol{J}_{\gamma\eta}\\ \boldsymbol{J}_{\eta\theta} & \boldsymbol{J}_{\eta\phi} & \boldsymbol{J}_{\eta\gamma} & \boldsymbol{J}_{\eta\eta}\end{bmatrix} \tag{3.34}$$

根据参数估计的原理知道 FIM 中的 $\boldsymbol{J}_{hk}\,(\boldsymbol{h},\boldsymbol{k}=\boldsymbol{\theta},\boldsymbol{\phi},\boldsymbol{\gamma},\boldsymbol{\eta})$ 等于:

$$\boldsymbol{J}_{hk}(i,j)=P\,\mathrm{Tr}\left(\boldsymbol{R}^{-1}\frac{\partial\boldsymbol{R}}{\partial\boldsymbol{h}_i}\boldsymbol{R}^{-1}\frac{\partial\boldsymbol{R}}{\partial\boldsymbol{k}_j}\right) \tag{3.35}$$

其中,P 是快拍个数,$\boldsymbol{R}$ 是接收数据的协方差矩阵,可通过如下计算得到:

$$\boldsymbol{R}=\sum_{k=1}^{K}\sigma_{sk}^2\boldsymbol{\alpha}(\theta_{z,k},\phi_{x,k},\gamma_k,\eta_k)\boldsymbol{\alpha}\ (\theta_{z,k},\phi_{x,k},\gamma_k,\eta_k)^{\mathrm{H}}+\sigma_n^2\boldsymbol{I} \tag{3.36}$$

其中,σ_{sk}^2 和 σ_n^2 是信号和噪声的功率。$\boldsymbol{R}$ 对未知的四个参数求偏导等于:

$$\frac{\partial\boldsymbol{R}}{\partial\theta_{x,k}}=\frac{\sigma_{sk}^2\,\partial\boldsymbol{\alpha}_k\boldsymbol{\alpha}_k^{\mathrm{H}}}{\partial\theta_{x,k}}=\sigma_{sk}^2\frac{\partial\boldsymbol{\alpha}_k}{\partial\theta_{x,k}}\boldsymbol{\alpha}_k^{\mathrm{H}}+\sigma_{sk}^2\boldsymbol{\alpha}_k\frac{\partial\boldsymbol{\alpha}_k^{\mathrm{H}}}{\partial\theta_{x,k}} \tag{3.37}$$

$$\frac{\partial\boldsymbol{R}}{\partial\phi_{x,k}}=\frac{\sigma_{sk}^2\,\partial\boldsymbol{\alpha}_k\boldsymbol{\alpha}_k^{\mathrm{H}}}{\partial\phi_{x,k}}=\sigma_{sk}^2\frac{\partial\boldsymbol{\alpha}_k}{\partial\phi_{x,k}}\boldsymbol{\alpha}_k^{\mathrm{H}}+\sigma_{sk}^2\boldsymbol{\alpha}_k\frac{\partial\boldsymbol{\alpha}_k^{\mathrm{H}}}{\partial\phi_{x,k}} \tag{3.38}$$

$$\frac{\partial\boldsymbol{R}}{\partial\gamma_k}=\frac{\sigma_{sk}^2\,\partial\boldsymbol{\alpha}_k\boldsymbol{\alpha}_k^{\mathrm{H}}}{\partial\gamma_k}=\sigma_{sk}^2\frac{\partial\boldsymbol{\alpha}_k}{\partial\gamma_k}\boldsymbol{\alpha}_k^{\mathrm{H}}+\sigma_{sk}^2\boldsymbol{\alpha}_k\frac{\partial\boldsymbol{\alpha}_k^{\mathrm{H}}}{\partial\gamma_k} \tag{3.39}$$

$$\frac{\partial\boldsymbol{R}}{\partial\eta_k}=\frac{\sigma_{sk}^2\,\partial\boldsymbol{\alpha}_k\boldsymbol{\alpha}_k^{\mathrm{H}}}{\partial\eta_k}=\sigma_{sk}^2\frac{\partial\boldsymbol{\alpha}_k}{\partial\eta_k}\boldsymbol{\alpha}_k^{\mathrm{H}}+\sigma_{sk}^2\boldsymbol{\alpha}_k\frac{\partial\boldsymbol{\alpha}_k^{\mathrm{H}}}{\partial\eta_k} \tag{3.40}$$

下面详细计算$\dfrac{\partial\boldsymbol{\alpha}_k}{\partial\theta_{z,k}}$、$\dfrac{\partial\boldsymbol{\alpha}_k}{\partial\phi_{z,k}}$、$\dfrac{\partial\boldsymbol{\alpha}_k}{\partial\gamma_k}$和$\dfrac{\partial\boldsymbol{\alpha}_k}{\partial\eta_k}$。注意在下面的计算中为了方便省略标号 k,则有

$$\boldsymbol{\alpha}(\theta_z,\phi_x,\gamma,\eta)=a_x(\theta_z,\phi_x)\otimes[\boldsymbol{\alpha}_y(\theta_z,\phi_x)\odot\boldsymbol{\alpha}^{(L)}(\theta_z,\phi_x,\gamma,\eta)]\in\mathbb{C}^{8\times1} \tag{3.41}$$

定义符号:

$$\boldsymbol{\alpha}\triangleq\boldsymbol{\alpha}(\theta_z,\phi_x,\gamma,\eta) \tag{3.42}$$

$$\boldsymbol{\alpha}_x=\boldsymbol{\alpha}_x(\theta_z,\phi_x)=[1\ \mathrm{e}^{-\mathrm{j}\frac{2\pi}{\lambda}D_x\sin\theta_z\sin\phi_x}]^{\mathrm{T}} \tag{3.43}$$

$$\boldsymbol{\alpha}_y=\boldsymbol{\alpha}_y(\theta_z,\phi_x)=[\mathrm{e}^{-\mathrm{j}\frac{2\pi}{\lambda}y_a\nu}\ \mathrm{e}^{-\mathrm{j}\frac{2\pi}{\lambda}y_b\nu}\ \mathrm{e}^{-\mathrm{j}\frac{2\pi}{\lambda}y_c\nu}\ \mathrm{e}^{-\mathrm{j}\frac{2\pi}{\lambda}D_y\nu}]^{\mathrm{T}} \tag{3.44}$$

$$\boldsymbol{\alpha}(L)(\theta_z,\phi_x,\gamma,\eta)=\boldsymbol{e}\odot\boldsymbol{l}\odot\boldsymbol{c}=-\begin{bmatrix}e_x\\e_y\\e_z\\e_x\end{bmatrix}\odot\begin{bmatrix}\ell_{\theta_x}^L\\\ell_{\theta_y}^L\\\ell_{\theta_z}^L\\\ell_{\theta_x}^L\end{bmatrix}\odot\begin{bmatrix}\csc\theta_x\\\csc\theta_y\\\csc\theta_z\\\csc\theta_x\end{bmatrix} \tag{3.45}$$

通过上述定义将导向矢量转化为如下的简化形式：

$$\boldsymbol{\alpha}=\boldsymbol{\alpha}_x\otimes[\boldsymbol{\alpha}_y\odot\boldsymbol{e}\odot\boldsymbol{l}\odot\boldsymbol{c}] \tag{3.46}$$

通过上述简化形式，可以得到偏导的具体计算值：

$$\begin{aligned}\frac{\partial\boldsymbol{\alpha}_k}{\partial\theta_k}=&\bar{\boldsymbol{\alpha}}_{x,\theta}\otimes[\boldsymbol{\alpha}_y\odot\boldsymbol{e}\odot\boldsymbol{l}\odot\boldsymbol{c}]+\boldsymbol{\alpha}_x\otimes[\bar{\boldsymbol{\alpha}}_{y,\theta}\odot\boldsymbol{e}\odot\boldsymbol{l}\odot\boldsymbol{c}]+\\&\boldsymbol{\alpha}_x\otimes[\boldsymbol{\alpha}_y\odot\bar{\boldsymbol{e}}_\theta\odot\boldsymbol{l}\odot\boldsymbol{c}]+\boldsymbol{\alpha}_x\otimes[\boldsymbol{\alpha}_y\odot\boldsymbol{e}\odot\bar{\boldsymbol{l}}_{\boldsymbol{\theta}}\odot\boldsymbol{c}]+\\&\boldsymbol{\alpha}_x\otimes[\boldsymbol{\alpha}_y\odot\boldsymbol{e}\odot l\odot\bar{\boldsymbol{c}}_\theta]\end{aligned} \tag{3.47}$$

$$\bar{\boldsymbol{\alpha}}_{x,\theta}=\left[0\ -\mathrm{j}\frac{2\pi}{\lambda}D_x\cos\theta_z\sin\phi_x\right]^{\mathrm{T}}\odot\boldsymbol{\alpha}_x \tag{3.48}$$

$$\bar{\boldsymbol{\alpha}}_{y,\theta}=-\mathrm{j}\frac{2\pi}{\lambda}\sin\theta_z\sin\phi_x\ [y_a\ y_b\ y_c\ D_y]^{\mathrm{T}}\odot\alpha_y \tag{3.49}$$

$$\bar{\boldsymbol{e}}_\theta=\begin{bmatrix}-\sin\theta_z\cos\phi_x\sin\gamma\mathrm{e}^{\mathrm{j}\eta}\\-\sin\theta_z\sin\phi_x\sin\gamma\mathrm{e}^{\mathrm{j}\eta}\\-\cos\theta_z\sin\gamma\mathrm{e}^{\mathrm{j}\eta}\\-\sin\theta_z\cos\phi_x\sin\gamma\mathrm{e}^{\mathrm{j}\eta}\end{bmatrix} \tag{3.50}$$

$$\bar{\boldsymbol{c}}_\theta=\begin{bmatrix}\dfrac{\cos\theta_z\cos\phi_x}{\sin^2\theta_x\sqrt{1-\sin^2\theta_z\ \cos^2\phi_x}}\\\dfrac{-\cos\theta_z\sin\phi_x}{\sin^2\theta_y\sqrt{1-\sin^2\theta_z\ \sin^2\phi_x}}\\\dfrac{-1}{\sin^2\theta_z}\\\dfrac{\cos\theta_z\cos\phi_x}{\sin^2\theta_x\sqrt{1-\sin^2\theta_z\ \cos^2\phi_x}}\end{bmatrix} \tag{3.51}$$

$$\bar{\boldsymbol{l}}_{\boldsymbol{\theta}}=[l_{\theta1}\ l_{\theta2}\ l_{\theta3}\ l_{\theta1}]^{\mathrm{T}} \tag{3.52}$$

$$l_{\theta1}=\frac{-\cos\theta_z\cos\phi_x f_{\theta1}}{\pi\cos^2\theta_x\sin(\pi L/\lambda)\sqrt{1-\sin^2\theta_z\cos^2\phi_x}} \tag{3.53}$$

$$f_{\theta1}=\lambda\cos\theta_x[\cos(\pi L\cos\theta_x/\lambda)-\cos(\pi L/\lambda)]-\pi L\sin^2\theta_x\sin(\pi L\cos\theta_x/\lambda) \tag{3.54}$$

$$\theta_x=\arccos(\sin\theta_z\cos\phi_x) \tag{3.55}$$

$$l_{\theta2}=\frac{-\cos\theta_z\sin\phi_x f_{\theta2}}{\pi\cos^2\theta_y\sin(\pi L/\lambda)\sqrt{1-\sin^2\theta_z\cos^2\phi_x}} \tag{3.56}$$

$$f_{\theta2}=\pi L\sin^2\theta_y\sin(\pi L\cos\theta_y/\lambda)-\lambda\cos\theta_y[\cos(\pi L\cos\theta_y/\lambda)-\cos(\pi L/\lambda)] \tag{3.57}$$

$$\theta_y=\arcsin(\sin\theta_z\sin\phi_x) \tag{3.58}$$

$$l_{\theta3}=\frac{\lambda\cos\theta_z[\cos(\pi L\cos\theta_z/\lambda)-\cos(\pi L/\lambda)]-\pi L\sin^2\theta_z\sin(\pi L\cos\theta_z/\lambda)}{\pi\cos^2\theta_z\sin(\pi L/\lambda)} \tag{3.59}$$

$$\begin{aligned}\frac{\partial\boldsymbol{\alpha}}{\partial\phi_x}=&[\bar{\boldsymbol{\alpha}}_{x,\phi_x}]\otimes[\boldsymbol{\alpha}_y\odot\boldsymbol{e}\odot\boldsymbol{l}\odot\boldsymbol{c}]+\boldsymbol{\alpha}_x\otimes[\bar{\boldsymbol{\alpha}}_{y,\phi_x}\odot\boldsymbol{e}\odot\boldsymbol{l}\odot\boldsymbol{c}]+\\&\boldsymbol{\alpha}_x\otimes[\boldsymbol{\alpha}_y\odot\bar{\boldsymbol{\alpha}}_{\phi_x}\odot\boldsymbol{l}\odot\boldsymbol{c}]+\boldsymbol{\alpha}_x\otimes[\boldsymbol{\alpha}_y\odot\boldsymbol{e}\odot\bar{\boldsymbol{l}}_{\phi_x}\odot\boldsymbol{c}]+\\&\boldsymbol{\alpha}_x\otimes[\boldsymbol{\alpha}_y\odot\boldsymbol{e}\odot\boldsymbol{l}\odot\bar{\boldsymbol{\alpha}}_{\phi_x}]\end{aligned} \tag{3.60}$$

$$\bar{\boldsymbol{\alpha}}_{x,\phi_x}=\left[0\ \ -\mathrm{j}\frac{2\pi}{\lambda}D_x\sin\theta_z\cos\phi_x\right]^{\mathrm{T}}\odot\boldsymbol{\alpha}_x \tag{3.61}$$

$$\bar{\boldsymbol{\alpha}}_{y,\phi_x}=\mathrm{j}\frac{2\pi}{\lambda}\sin\theta_z\sin\phi_x\ [y_a\ y_b\ y_c\ D_y]^{\mathrm{T}}\odot\boldsymbol{\alpha}_y \tag{3.62}$$

$$\bar{\boldsymbol{e}}_{\phi_x}=\begin{bmatrix}-\cos\theta_z\sin\phi_x\sin\gamma e^{\mathrm{j}\eta}-\cos\phi_x\cos\gamma\\ \cos\theta_z\cos\phi_x\sin\gamma e^{\mathrm{j}\eta}-\sin\phi_x\cos\gamma\\ 0\\ -\cos\theta_z\sin\phi_x\sin\gamma e^{\mathrm{j}\eta}-\cos\phi_x\cos\gamma\end{bmatrix} \tag{3.63}$$

$$\bar{\boldsymbol{l}}_{\phi_x}=[l_{\phi1}\ l_{\phi2}\ l_{\phi3}\ l_{\phi1}]^{\mathrm{T}} \tag{3.64}$$

$$l_{\phi1}=\frac{-\sin\theta_z\sin\phi_x f_{\phi1}}{\pi\cos^2\theta_x\sin(\pi L/\lambda)\sqrt{1-\sin^2\theta_z\cos^2\phi_x}} \tag{3.65}$$

$$f_{\phi 1}=\pi L\ \sin^2\theta_x \sin(\pi L\cos\theta_x/\lambda)-\lambda\cos\theta_x[\cos(\pi L\cos\theta_x/\lambda)-\cos(\pi L/\lambda)] \tag{3.66}$$

$$l_{\phi 2}=\frac{-\sin\theta_z\cos\phi_x f_{\phi 2}}{\pi\cos^2\theta_y\sin(\pi L/\lambda)\sqrt{1-\sin^2\theta_z\ \cos^2\phi_x}} \tag{3.67}$$

$$f_{\phi 2}=\pi L\ \sin^2\theta_y \sin(\pi L\cos\theta_y/\lambda)-\lambda\cos\theta_y[\cos(\pi L\cos\theta_y/\lambda)-\cos(\pi L/\lambda)] \tag{3.68}$$

$$l_{\phi 3}=\frac{\lambda\cos\theta_z[\cos(\pi L\cos\theta_z/\lambda)-\cos(\pi L/\lambda)]-\pi L\ \sin^2\theta_z\sin(\pi L\cos\theta_z/\lambda)}{\pi\cos^2\theta_z\sin(\pi L/\lambda)} \tag{3.69}$$

$$\bar{\boldsymbol{c}}_{\phi x}=\begin{bmatrix}\dfrac{-\sin\theta_z\sin\phi_x}{\sin^2\theta_x\sqrt{1-\sin^2\theta_z\ \cos^2\phi_x}}\\ \dfrac{-\sin\theta_z\cos\phi_x}{\sin^2\theta_y\sqrt{1-\sin^2\theta_z\ \sin^2\phi_x}}\\ 0\\ \dfrac{-\sin\theta_z\sin\phi_x}{\sin^2\theta_x\sqrt{1-\sin^2\theta_z\ \cos^2\phi_x}}\end{bmatrix} \tag{3.70}$$

$$\frac{\partial\boldsymbol{\alpha}}{\partial\gamma}=\boldsymbol{\alpha}_x\otimes[\boldsymbol{\alpha}_y\odot\bar{\boldsymbol{e}}_\gamma\odot\boldsymbol{l}\odot\boldsymbol{c}] \tag{3.71}$$

$$\bar{\boldsymbol{e}}_\gamma=\begin{bmatrix}\cos\theta_z\cos\phi_x\cos\gamma\mathrm{e}^{\mathrm{j}\eta}+\sin\phi_x\sin\gamma\\ \cos\theta_z\sin\phi_x\cos\gamma\mathrm{e}^{\mathrm{j}\eta}-\cos\phi_x\sin\gamma\\ -\sin\theta_z\cos\gamma\mathrm{e}^{\mathrm{j}\eta}\\ \cos\theta_z\cos\phi_x\cos\gamma\mathrm{e}^{\mathrm{j}\eta}+\sin\phi_x\sin\gamma\end{bmatrix} \tag{3.72}$$

$$\frac{\partial\boldsymbol{\alpha}}{\partial\eta}=\boldsymbol{\alpha}_x\otimes[\boldsymbol{\alpha}_y\odot\bar{\boldsymbol{e}}_\eta\odot\boldsymbol{l}\odot\boldsymbol{c}] \tag{3.73}$$

$$\bar{\boldsymbol{e}}_\eta=\begin{bmatrix}j\cos\theta_z\cos\phi_x\sin\gamma_k\mathrm{e}^{\mathrm{j}\eta}\\ j\cos\theta_z\sin\phi_x\sin\gamma_k\mathrm{e}^{\mathrm{j}\eta}\\ -j\sin\theta_z\sin\gamma\mathrm{e}^{\mathrm{j}\eta}\\ j\cos\theta_z\cos\phi_x\sin\gamma_k\mathrm{e}^{\mathrm{j}\eta}\end{bmatrix} \tag{3.74}$$

§3.4 仿真结果分析

本节用计算机仿真来验证算法的有效性。假设子阵的前三个正交阵元的位置配置为：$y_a=0$，$y_b=0.5\lambda$，$y_c=\lambda$。x 轴中子阵之间的距离 $D_x=5(\lambda/2)$，y 轴中第一个天线元素和第四个元素的间距为 $D_y=5(\lambda/2)$。八个阵元的位置均已确定。假设同一个距离单元有两个($K=2$)独立目标，$(\theta_1,\phi_1,\gamma_1,\eta_1)=(10°,15°,45°,-90°)$，$(\theta_2,\phi_2,\gamma_2,\eta_2)=(60°,35°,45°,90°)$。可看出两个信号分别为右旋圆极化信号和左旋圆极化信号。

仿真一：目标二维 DOA 估计及其配对。

剩下的两个比较重要的参数是信噪比和快拍数，设置为 SNR＝20 dB 和 $L=200$。将 100 次独立实验的二维 DOA 估计结果显示在图 3.6 中。从图3.6 可以看出，二维 DOA 的估计结果与仿真中的预设值一致，且多目标的二维 DOA 配对结果正确，初步验证了算法的正确性。下面做深入的仿真，以进一步验证算法的有效性。

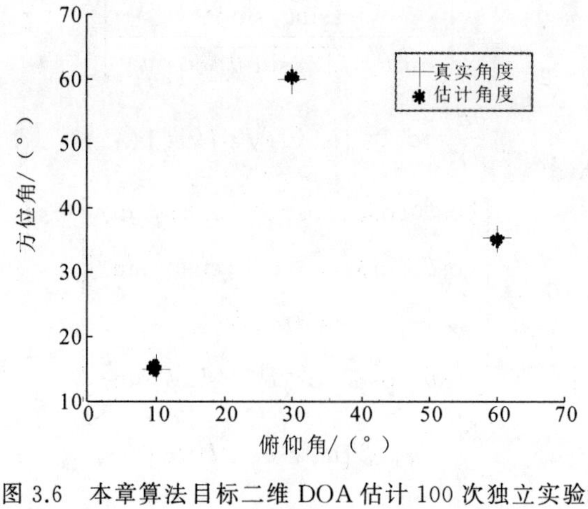

图 3.6　本章算法目标二维 DOA 估计 100 次独立实验结果

仿真二：RMSE 随信噪比变化曲线。

设置快拍数为 $L=200$，对信噪比进行变化处理。每个数据点做 1 000 次独立实验。对俯仰角和方位角进行估计，分别统计其 RMSE，其中 RMSE 的定义与第二章中的定义类似，这里不再赘述。从图 3.7 中可以看到两个目标的二维 DOA 的 RMSE 随着信噪比增大而减小，即估计性能越好，且接近于

RCRB，验证了算法的有效性。

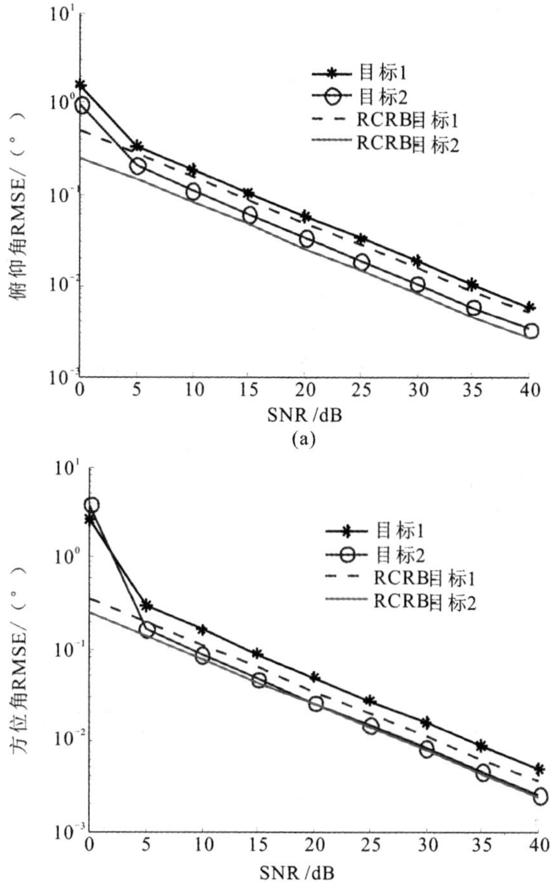

图 3.7　两维 DOA 估计 RMSE 与信噪比关系图
(a)俯仰角估计；(b)方位角估计

仿真三：估计性能与阵元大小的变化曲线。

除了设置两个重要参数的信噪比 SNR＝10 dB 和快拍数 $L=200$ 外，还设置阵元长度从 0.1 个波长变化到 0.9 个波长。每个数据点做 1 000 次独立实验，图 3.8 给出了二维 DOA 估计 RMSE 与阵元长度大小的关系图。从图 3.8 中可看出，方位角和俯仰角的最终估计值与电偶极子的长度并没有直接关系，因为二维方向的有效孔径固定。但是另一方面，方位角和俯仰角的粗估计精度是随着电偶极子长度的增加而提高的。

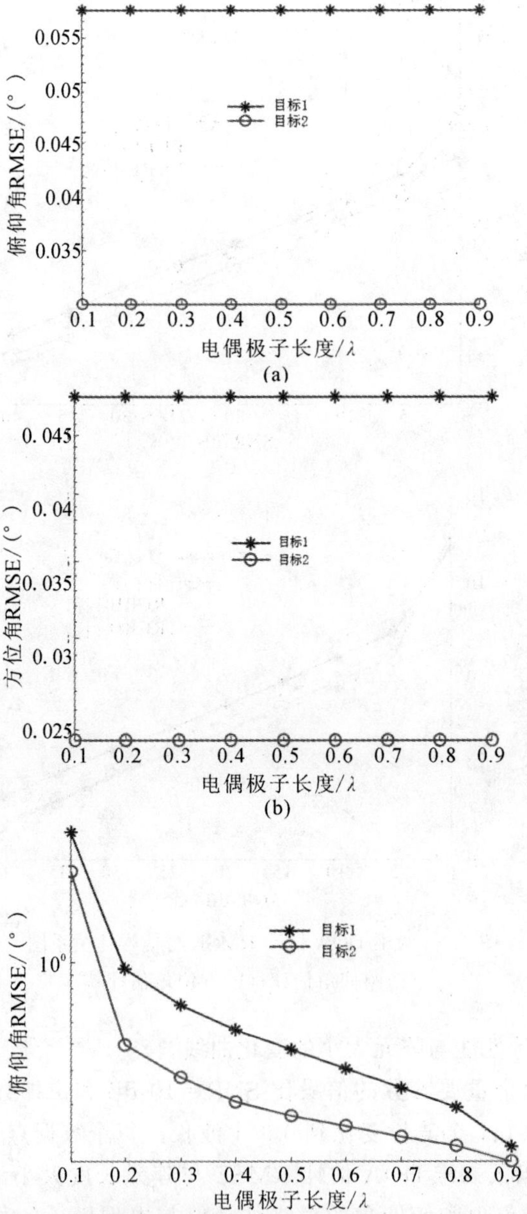

图 3.8　二维 DOA 估计 RMSE 与阵元大小的长度变化关系分析

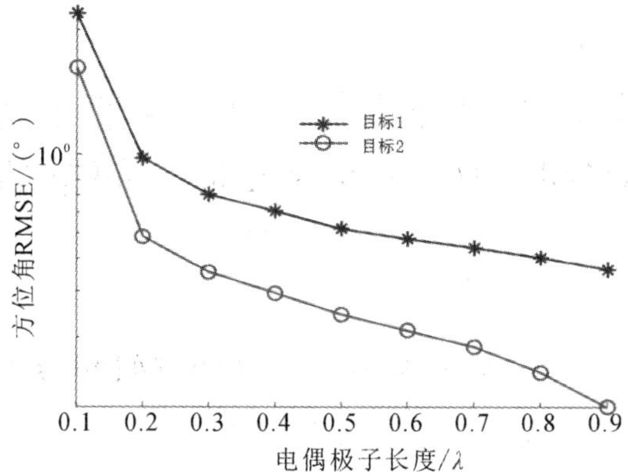

(d)

续图 3.8　二维 DOA 估计 RMSE 与阵元大小的长度变化关系分析

(a)俯仰角无模糊精估计值;(b)方位角无模糊精估计值;

(c)俯仰角无模糊粗估计值;(d)方位角无模糊粗估计值

§3.5　本 章 小 结

本章研究了“长”电偶极子和“大”磁环组成的阵列的二维 DOA 的估计问题。二维的稀疏阵列配置可达到阵元数小、二维孔径大的效果,同时大大缓解了共点式极化矢量天线互耦严重的问题。更为重要的是本章采用的电偶极子的长度较长,其辐射效率比短电偶极子要大得多,为工程实验奠定了重要的理论基础。

本章设计的极化矢量阵列与经典极化矢量阵列的对比见表 3.1。

表 3.1　本章设计的极化矢量阵列与经典极化矢量阵列对比

项　目	短电偶极子共点式矢量传感器	短电偶极子分离式矢量传感器	经典长电偶极子分离式矢量传感器	本章长电偶极子分离式矢量传感器
互耦	强	弱	弱	弱
辐射效率	低	低	高	高
先验信息	否	否	是	否

第4章　基于波形分集的FDA与速度矢量传感器MIMO阵列多参数估计问题研究

§4.1　基于参差频率的FDA MIMO阵列的角度和距离联合估计

§4.1.1　引言

对于角度和距离的联合估计是FDA-MIMO雷达的一个重要研究课题。当发射阵列均匀加权时，或者说是没有对发射方向图进行特别的优化设计，那么其发射方向图函数为[222]

$$|AF(r,\theta)|=\frac{\sin\left[M\pi\left(-\frac{\Delta f r}{c}+\frac{f_0 d\sin\theta}{c}+\frac{\Delta f d\sin\theta}{c}\right)\right]}{\sin\left[\pi\left(-\frac{\Delta f r}{c}+\frac{f_0 d\sin\theta}{c}+\frac{\Delta f d\sin\theta}{c}\right)\right]}$$

这是一个辛克函数，显然其在$\frac{\Delta f r}{c}+\frac{d f_0\sin\theta}{c}+\frac{d\Delta f\sin\theta}{c}=k,k=0,\pm1,\pm2,\cdots$处会出现栅瓣，其中，$d$表示阵元间距，$\theta$表示目标DOA，$r$为目标距离，$M$为阵元个数，$f_0$、$\Delta f$分别表示初始载频和频率增量。对于FDA-MIMO雷达来说，在估计DOA时，受到接收阵列方向的约束，使得在估计DOA时不会出现模糊，但是距离维上的栅瓣仍然难以避免，即在估计目标距离时会出现周期性模糊。距离分辨力等于$\frac{c}{2(M-1)\Delta f}$[225]。在某些情况下，需要较高的距离分辨力，在阵列大小确定的前提下，即M固定，则需要提高频率增量Δf，这时方向图所导致的距离模糊度变高。举例来说，对于米波雷达来说要求150 m的距离分辨力是合理的。那么对于21个天线的阵列来说需要1 MHz带宽，频率增量等于50 000 Hz，设$\theta=0°$，则在$\frac{\Delta f r}{c}=k,k=0,\pm1,\pm2,\ldots$会出现模糊值，在间隔6 km距离时会出现周期性模糊，这是雷达设计者难以接受的。受到参差MTI解距离模糊思想的启发，本书提出了采用参差频率增量来扩大距离估计的模糊范围,具体来说就是,通过发射不一样的频率增量的脉冲来解决距离模糊问题。

FDA-MIMO 雷达的角度和距离联合估计对于大多数算法来说都需要在距离维和角度维进行二维联合搜索,文献[225]则提出了一种利用发射双脉冲的方法来进行两个一维搜索实现角度和距离的联合估计,但是其并没有解决距离模糊的问题,都是搜索类算法,计算量大。鉴于此,本章提出一种 ESPRIT 和 MUSIC 算法相结合的办法,具体是先采用 ESPRIT 算法估计目标角度,然后利用估计的角度值采用 MUSIC 超分辨算法进行一维距离搜索得到距离值,该方法的计算量比文献[225]要小得多,且目标角度和距离估计值是自动配对的,同时参差频率增量扩大了距离模糊范围。总体来说,本章的贡献主要有两点,一是利用参差频率增量来扩大距离估计的模糊范围,二是采用 ESPRIT 和 MUSIC 联合估计算法来减少计算量。

§ 4.1.2　FDA-MIMO 雷达参数估计信号模型

M 个阵元组成的 FDA,其每个阵元的频率为

$$f_m = f_0 + m\Delta f,\quad m = 0 \cdots M-1 \tag{4.1}$$

其中,f_0 表示阵列参考单元的载频,Δf 表示相邻阵元间的频率增量。那么根据文献[225],可以知道发射导向矢量为

$$\boldsymbol{\alpha}_t(r,\theta) = \left\{1 \cdots \exp\left[-\mathrm{j}2\pi(M-1)\left(-\frac{\Delta f r}{c} + \frac{f_0 d\sin\theta}{c} + \frac{\Delta f d\sin\theta}{c}\right)\right]\right\} \tag{4.2}$$

其中,r 表示目标的斜距,下面直接用距离表示,θ 表示目标的角度,c 表示光速,是一个常量。发射导向矢量是角度和距离二维依赖的。假设该 MIMO 雷达是收发共置的,则接收阵列角度的导向矢量为

$$\boldsymbol{\alpha}_r(\theta) = [1\ \exp(-\mathrm{j}2\pi d\sin\theta/\lambda) \cdots \exp(-\mathrm{j}2\pi d\sin\theta(M-1)/\lambda)]^{\mathrm{T}} \tag{4.3}$$

则 FDA-MIMO 雷达的导向矢量可以写为

$$\boldsymbol{\alpha}(r,\theta) = \boldsymbol{\alpha}_t(r,\theta) \otimes \boldsymbol{\alpha}_r(\theta) \tag{4.4}$$

则 FDA-MIMO 雷达的接收数据可以表示为

$$x(t) = \sum_{k=1}^{K} \boldsymbol{\alpha}(r_k,\theta_k)s_k(t) + \boldsymbol{n}(t) = \mathbf{A}\boldsymbol{s}(t) + \boldsymbol{n}(t) \tag{4.5}$$

其中,K 为目标个数。导向矩阵 $\mathbf{A} = [\boldsymbol{\alpha}(r_1,\theta_1) \cdots \boldsymbol{\alpha}(r_K,\theta_K)] \in \mathbb{C}^{M^2 \times K}$,信号矢量等于 $\boldsymbol{s}(t) = [s_1(t) \cdots s_K(t)]^{\mathrm{T}} \in \mathbb{C}^{K\times 1}$。噪声 $\boldsymbol{n}(t)$ 设置为高斯白噪声。那么接下来的目的就是根据式(4.5)计算出 K 个目标的角度和距离值。

§4.1.3 参差频率增量的 ESPRIT 和 MUSIC 联合估计方法

1.参差频率增量解距离模糊

FDA-MIMO 雷达在距离维上是有模糊的，其模糊值为$\frac{\Delta f r}{c}+\frac{d f_0 \sin\theta}{c}+\frac{d\Delta f \sin\theta}{c}=k, k=0,\pm1,\pm2,\ldots$。现在利用与参差周期解 MTI 的高重频所带来的距离模糊一样，在脉冲之间发射参差的频率增量，为了方便说明，给出两个参差频率增量的说明如下：

$$\left.\begin{aligned}\frac{\Delta f_1 r_{\text{amb1}}}{c}+\frac{d f_0 \sin\theta}{c}+\frac{d\Delta f_1 \sin\theta}{c}=k_1,\quad k_1=0,\pm1,\pm2,\ldots\\ \frac{\Delta f_2 r_{\text{amb2}}}{c}+\frac{d f_0 \sin\theta}{c}+\frac{d\Delta f_2 \sin\theta}{c}=k_2,\quad k_2=0,\pm1,\pm2,\ldots\end{aligned}\right\}\tag{4.6}$$

其中，Δf_1 是第一个发射脉冲的频率增量，Δf_2 是第二个发射脉冲的频率增量，r_{amb1} 表示 Δf_1 所对应的最大距离不模糊值，r_{amb2} 表示 Δf_2 所对应的最大距离不模糊值。那么发射两个脉冲的最大不模糊距离则为 r_{amb1} 和 r_{amb2} 的重叠部分，即 $r_{\text{amb}}=\min \text{amb}\{r_{\text{amb1}}, r_{\text{amb2}}\}$，其中 $\min \text{amb}\{x,y\}$ 表示取集合 x 和集合 y 相同值的最小部分。那么，只要合理设计 $\Delta f_1, \Delta f_2$ 就可以使距离模糊变成设计者想要的模糊距离值。

这里作者提出一种模糊距离互质的方法，式(4.6)中有未知量 θ 的影响，很难进行分析，可将 θ 设计成一个定值进行简化，为了方便说明设置 $\theta=0°$，则式(4.6)简化成如下形式：

$$\left.\begin{aligned}\frac{\Delta f_1 r_{\text{amb1}}}{c}=k_1, k_1=0,\pm1,\pm2,\ldots\\ \frac{\Delta f_2 r_{\text{amb2}}}{c}=k_2, k_2=0,\pm1,\pm2,\ldots\end{aligned}\right\}\tag{4.7}$$

对式(4.7)求解得到：

$$r_{\text{amb}}=\min \text{amb}\{r_{\text{amb1}}, r_{\text{amb2}}\}=\min \text{amb}\left\{\frac{ck_1}{\Delta f_1}, \frac{ck_2}{\Delta f_2}\right\}\tag{4.8}$$

对于式(4.8)，令 $k_1=k_2=1$，且选取$\frac{c}{\Delta f_1}$与$\frac{c}{\Delta f_2}$的比值为互质关系，即参差比为互质关系。则最大模糊距离等于：

$$r_{\text{amb}}=\min \text{amb}\left\{\frac{c}{\Delta f_1}, \frac{c}{\Delta f_2}\right\}=\frac{c}{\Delta f_1}\frac{c}{\Delta f_2}\tag{4.9}$$

2.基于 ESPRIT 和 MUSIC 的角度和距离联合估计

要想利用 ESPRIT 算法，就要挖掘接收数据的旋转不变性。由于受到距离因素的影响，对于整个导向矢量 $\boldsymbol{\alpha}(r,\theta)$ 来说，很难找到它的旋转不变性。但是对于接收导向矢量 $\boldsymbol{\alpha}_r(\theta)$ 来说，若它是一个 ULA 的导向矢量，很容易找到空域旋转不变性，然后将其推广至整个导向矢量。下面进行具体说明。定义 $\boldsymbol{\alpha}_1(r,\theta)\triangleq\boldsymbol{\alpha}_t(r,\theta)\otimes\boldsymbol{\alpha}_{r,1}(\theta)$，其中 $\boldsymbol{\alpha}_{r,1}(\theta)$ 表示前面 $(M-1)$ 个接收阵元的导向矢量：

$$\boldsymbol{\alpha}_{r,1}(\theta)\triangleq[1\ \mathrm{e}^{-\mathrm{j}2\pi d\sin\theta/\lambda}\cdots\ \mathrm{e}^{-\mathrm{j}2\pi d\sin\theta(M-2)/\lambda}]^{\mathrm{T}} \tag{4.10}$$

定义 $\boldsymbol{\alpha}_2(r,\theta)\triangleq\boldsymbol{\alpha}_t(r,\theta)\otimes\boldsymbol{\alpha}_{r,2}(\theta)$，其中 $\boldsymbol{\alpha}_{r,2}(\theta)$ 表示后面的 $(M-1)$ 个接收阵元的导向矢量：

$$\boldsymbol{\alpha}_{r,2}(\theta)\triangleq[\mathrm{e}^{-\mathrm{j}2\pi d\sin\theta/\lambda}\cdots\ \mathrm{e}^{-\mathrm{j}2\pi d\sin\theta(M-1)/\lambda}]^{\mathrm{T}} \tag{4.11}$$

这样对于 $\boldsymbol{\alpha}_1(r,\theta)$ 和 $\boldsymbol{\alpha}_2(r,\theta)$，有如下的旋转不变关系：

$$\begin{aligned}\boldsymbol{\alpha}_2(r,\theta)&=\boldsymbol{\alpha}_t(r,\theta)\otimes\boldsymbol{\alpha}_{r,2}(\theta)\\&=\boldsymbol{\alpha}_t(r,\theta)\otimes[\boldsymbol{\alpha}_{r,1}(\theta)\mathrm{e}^{-\mathrm{j}2\pi d\sin\theta/\lambda}]\\&=\boldsymbol{\alpha}_1(r,\theta)\mathrm{e}^{-\mathrm{j}2\pi d\sin\theta/\lambda}\end{aligned} \tag{4.12}$$

式(4.12)可转化为：

$$\boldsymbol{J}_2\boldsymbol{\alpha}(r,\theta)=\mathrm{e}^{-\mathrm{j}2\pi d\sin\theta/\lambda}\boldsymbol{J}_1\boldsymbol{\alpha}(r,\theta) \tag{4.13}$$

其中，选择矩阵 $\boldsymbol{J}_1=\boldsymbol{I}_M\otimes[\boldsymbol{I}_M\ \boldsymbol{0}_{(M-1)\times1}]$，$\boldsymbol{J}_2=\boldsymbol{I}_M\otimes[\boldsymbol{0}_{(M-1)\times1}\ \boldsymbol{I}_M]$。式(4.13)可得到关于整个导向矢量的旋转不变性。对于导向矩阵来说，则有

$$\left.\begin{aligned}&\boldsymbol{J}_2\boldsymbol{A}(r,\theta)=\boldsymbol{J}_1\boldsymbol{A}(r,\theta)\boldsymbol{\Theta}(\theta)\\&\boldsymbol{\Theta}(\theta)\triangleq\mathrm{diag}\,[\exp(-\mathrm{j}2\pi d\sin\theta_1/\lambda)\ \cdots\ \exp(-\mathrm{j}2\pi d\sin\theta_K/\lambda)]\end{aligned}\right\} \tag{4.14}$$

矩阵 $\boldsymbol{\Theta}(\theta)$ 只含有 DOA 信息，则可根据式(4.14)的旋转不变关系计算出 DOA 估计值。根据式(4.5)的接收数据，采用第二章中的分析方法得到信号子空间 $\boldsymbol{E}_S$ 和噪声子空间 $\boldsymbol{E}_n$，以及唯一非奇异矩阵 $\boldsymbol{T}$。将其代入式(4.14)得到

$$\boldsymbol{J}_2\boldsymbol{E}_S=\boldsymbol{J}_1\boldsymbol{E}_S\boldsymbol{\Psi}(\theta) \tag{4.15}$$

其中，$\boldsymbol{\Psi}(\theta)=\boldsymbol{T}^{-1}\boldsymbol{\Phi}(\theta)\boldsymbol{T}$。在噪声条件下，方程式(4.8)可用总体最小二乘方法求解得到 $\boldsymbol{\Psi}(\theta)$。故 DOA 估计值 $\hat{\theta}$ 可由下式计算得到，即

$$\hat{\theta}_k=-\arcsin\left\{\frac{\lambda}{2\pi d}\angle([\boldsymbol{\Phi}(\theta)]_{kk})\right\},\quad k=1,\cdots,K \tag{4.16}$$

其中，矩阵 $\boldsymbol{\Phi}(\theta)$ 的对角元素 $[\boldsymbol{\Phi}(\theta)]_{kk}$ 等于 $\boldsymbol{\Psi}(\theta)$ 特征分解之后的特征值。

根据信号模型可知，该阵列的 DOA 与距离参数可通过式(4.17)的二维常规 MUSIC 搜索算法来实现，即

$$f_{\text{2D-MUSIC}}(\theta,r)=\frac{1}{\boldsymbol{\alpha}(\theta,r)^{\mathrm{H}}\boldsymbol{E}_N\boldsymbol{E}_N^{\mathrm{H}}\boldsymbol{\alpha}(\theta,r)} \tag{4.17}$$

利用 ESPRIT 方法得到角度估计值 $\hat{\theta}_k, k=1,\cdots,K$ 后，将每个估计值分别带入式(4.17)中，求得同一个目标所对应的距离值，即

$$\hat{r}_k=\max\left[\frac{1}{\boldsymbol{\alpha}(\hat{\theta}_k,r_k)^{\mathrm{H}}\boldsymbol{E}_N\boldsymbol{E}_N^{\mathrm{H}}\boldsymbol{\alpha}(\hat{\theta}_k,r_k)}\right],\quad k=1,\cdots,K \tag{4.18}$$

式(4.18)所求的第 k 个目标的距离与其同一个目标的角度是一一对应的，即可自动配对。需要注意的是距离搜索范围为前面计算出的无模糊距离范围。该一维搜索方法与文献[225]一样，需要 K 次连续的一维搜索。

根据上述分析算法流程及算法计算量，关于计算量的比较，本书主要与文献[225]作对比。

(1)采用参差频率增量的 ESPRIT - MUSIC 算法的步骤和计算量：

第一步：形成数据 $\boldsymbol{x}=[\boldsymbol{x}_1\ \boldsymbol{x}_2]\in\mathbb{C}^{2M\times1}$，计算其协方差矩阵 $\boldsymbol{R}_{xx}=E[\boldsymbol{x}*\boldsymbol{x}]\in\mathbb{C}^{2M\times2M}$（$[\boldsymbol{x}_1,\boldsymbol{x}_2]$分别为两个脉冲的数据）。$\boldsymbol{R}_{xx}$ 特征值分解得到信号和噪声子空间，其计算量为 $O\{4M^4L+8M^6\}$。

第二步：根据信号子空间，利用 ESPRIT 计算 DOA 角度，得到 K 个目标的 DOA 估计值，其计算量为 $O\{6M(M-1)K^2+K^3\}$。

第三步：根据噪声子空间，将每个估计值代入式(4.18)计算其距离，其计算量为 $n[(8M^3+4M^2)(4M^2-K)+4M^2]$，$n$ 表示距离搜索次数。

(2)不采用参差频率增量的 ESPRIT - MUSIC 算法和计算量：

第一步：形成数据 $\boldsymbol{x}\in\mathbb{C}^{M\times1}$，计算其协方差矩阵 $\boldsymbol{R}_{xx}=E[\boldsymbol{x}*\boldsymbol{x}]\in\mathbb{C}^{M\times M}$（$\boldsymbol{x}\in\mathbb{C}^{M\times1}$ 为单个脉冲的数据）。$\boldsymbol{R}_{xx}$ 特征值分解得到信号和噪声子空间，其计算量为 $O\{(MM)^2L+(MM)^3\}$。

第二步：根据信号子空间，利用 ESPRIT 计算 DOA 角度，得到 K 个目标的 DOA 估计值，其计算量为 $O\{3M(M-1)K^2+K^3\}$。

第三步：根据噪声子空间，将每个估计值代入式(4.18)计算其距离，其计算量为 $n[(M^3+M^2)(M^2-K)+M^2]$，n 表示距离搜索次数。

(3)文献[225]算法的步骤和计算量：

第一步：数据 $\boldsymbol{x}_1\in\mathbb{C}^{M\times1}$，$\boldsymbol{x}_2\in\mathbb{C}^{M\times1}$，计算其协方差矩阵 $\boldsymbol{R}_{xx1}=E[\boldsymbol{x}_1*\boldsymbol{x}_1]\in\mathbb{C}^{M\times M}$，$\boldsymbol{R}_{xx2}=E[\boldsymbol{x}_2*\boldsymbol{x}_2]\in\mathbb{C}^{M\times M}$（$[\boldsymbol{x}_1,\boldsymbol{x}_2]$分别为两个脉冲的数据）。$\boldsymbol{R}_{xx1}$ 特征值分解得到噪声子空间 1，$\boldsymbol{R}_{xx2}$ 特征值分解得到噪声子空间 2，其计算量为 $O\{2M^4L+2M^6\}$。

第二步：根据噪声子空间 1，利用 MUSIC 搜索计算，得到 K 个目标的 DOA 估计值，其计算量为 $n_1[(M^3+M^2)(M^2-K)+M^2]$，$n_1$ 表示角度搜索次数。

第三步：根据噪声子空间 2，将每个估计值代入式(4.18)计算其距离，其计算量为 $n_2[(M^3+M^2)(M^2-K)+M^2]$，$n_2$ 表示距离搜索次数。

从上面的分析可以看出，只利用本章提出的 ESPRIT - MUSIC 算法肯定要比文献[225]计算量小，因为第一步和第三步计算量一样，本章第二步算法采用 ESPRIT，无需搜索，计算量比 MUSIC 算法要小得多。但该方法距离模糊与文献[225]一样。利用本章提出的参差频率增量的 ESPRIT - MUSIC 算法，其计算量与文献[225]相比，主要由阵元 M 和角度搜索次数 n_1 来决定，哪个占支配位置，哪个计算量就占优，但该方法的模糊距离范围肯定比文献[225]大。

本章所提算法需要注意的地方有：一是本算法不考虑高脉冲重复频率所带来的距离模糊，若需要可采用文献[226]的方法来解决此问题；二是本算法在计算角度时并没有利用整个阵列的孔径，只利用接收阵列的孔径，文献[225]则利用了全孔径；三是为了保证不模糊，选取阵列中相邻阵元间隔的时候，选取大的频率增量作为阵列的间隔，即

$$d=\frac{c}{2[f+(M-1)\max(\Delta f_1,\Delta f_2)]} \tag{4.19}$$

§4.1.4　仿真结果分析

假设阵元数为 $M=10$ 个，同一距离单元存在 $K=3$ 个独立目标，目标 DOA 为 $\theta=\{10°,20°,30°\}$，目标距离为 $r=\{10\ \text{km},20\ \text{km},30\ \text{km}\}$，快拍数 $L=200$，信噪比 SNR＝20 dB，两个互质的参差频率设置为 10 000 Hz 和 7 500 Hz。

如果不采用参差频率的话，那么两个频率增量所对应的距离模糊分别为 30 km 和 40 km。利用式(4.18)给出第一个目标即 $r=10$ km 估计的距离谱。图 4.1(a)(b)(c)分别给出了频率增量为 10 000 Hz、7 500 Hz 以及参差频率的结果。图 4.1(a)的距离估计结果为 10 km(真实目标距离)、40 km、70 km、100 km、130 km。图 4.1(b)的距离估计结果为 10 km(真实目标距离)、50 km、90 km、130 km。图 4.1(c)的距离估计结果为 10 km(真实目标距离)、130 km。可以看到图 4.1(c)的距离模糊值周期为 120 km。这与理论分析的结果是完全一致的。

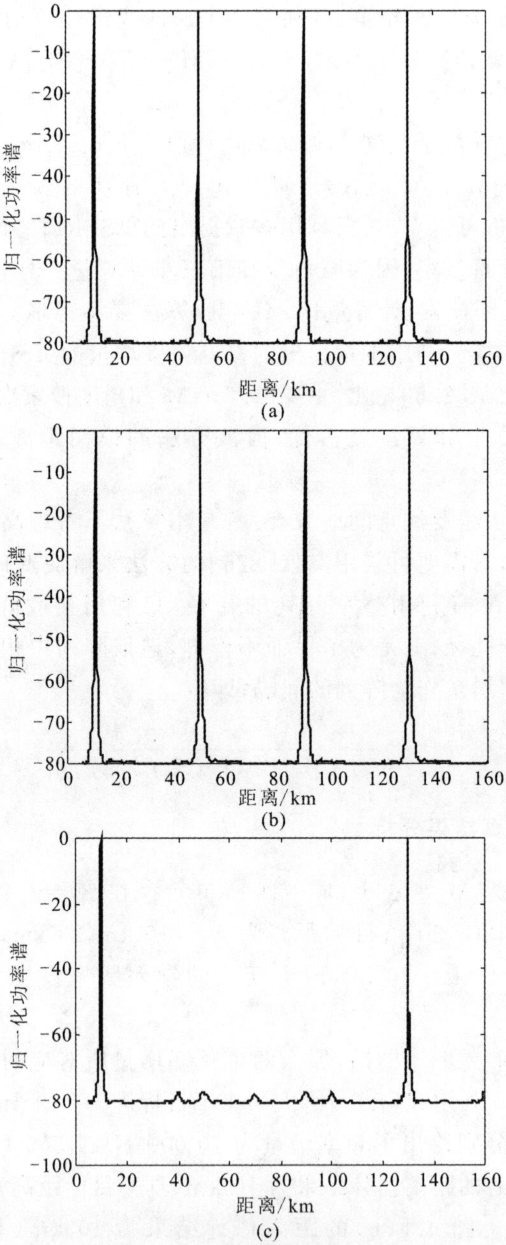

图 4.1　距离维搜索的估计结果

(a)频率增量为 10 000 Hz 的第一个目标的距离谱;(b)频率增量为 7 500 Hz 的第一个目标的距离谱;(c)参差频率增量第一个目标的距离谱

下面给出三个目标的角度和距离自动配对结果，这里所说的结果是在无模糊范围以内的。从图 4.2 中可以看到进行 100 次的蒙特卡洛实验的结果配对均正确。

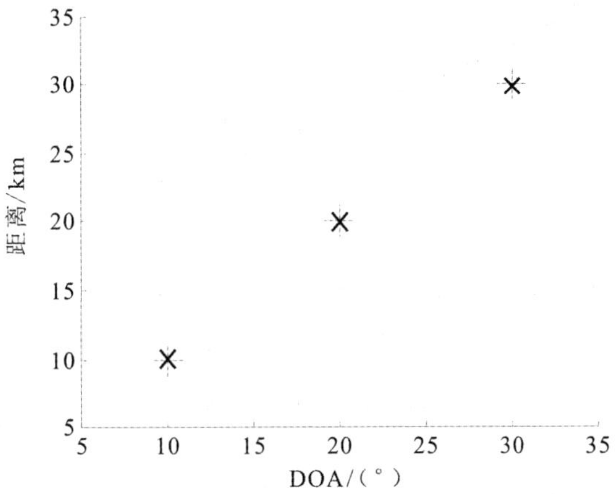

图 4.2　三个目标的角度和距离自动配对结果

图 4.3 所示为文献[225]的角度估计和距离估计的 RMSE 随信噪比变化曲线的两个对比图。为了对比的公平性，本书将文献[225]中采用的 DBF 算法改成超分辨算法，即角度和距离都采用一维搜索的 MUSIC 算法来进行对比。在多目标的前提下，MUSIC 算法是要优于 DBF 算法的。

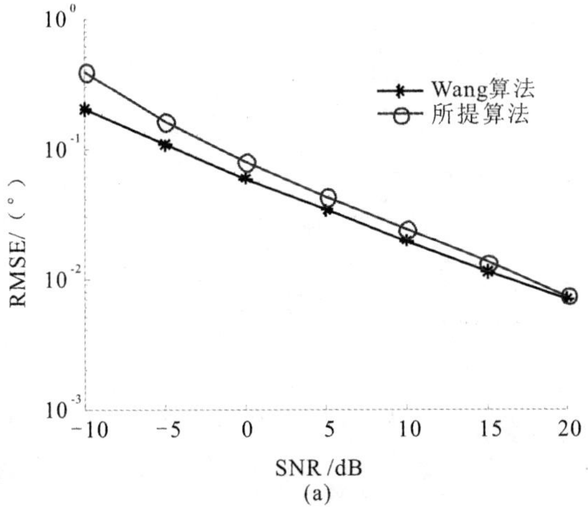

图 4.3　RMSE 估计结果

(a)角度估计结果；

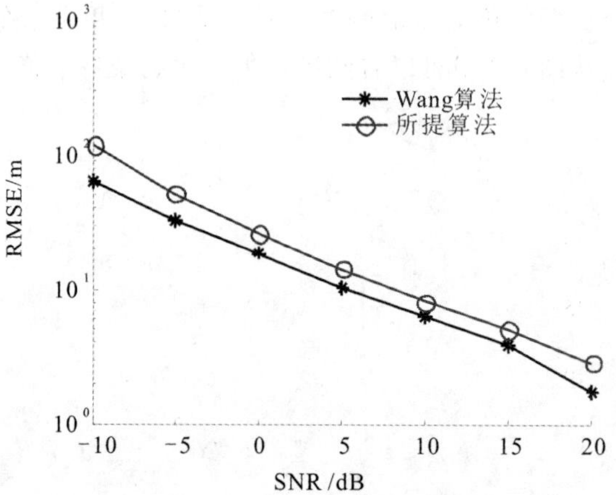

(b)

续图 4.3　RMSE 估计结果

(b)距离估计结果

在计算运行时间对比图中(见图 4.4),设置搜索的距离范围均为 5～35 km,搜索间隔为 5 m,本书算法则采用两个脉冲、单个脉冲的方法来与文献[225]比较计算量。文献[225]中的角度搜索范围为－90°～90°,间隔设为三种:0.1°、0.01°、0.001°。可以看出所提算法若采用单个脉冲则计算量要比文献[225]小,采用两个脉冲则计算量大于 0.1°间隔和小于 0.001°间隔,与0.01°间隔相当。

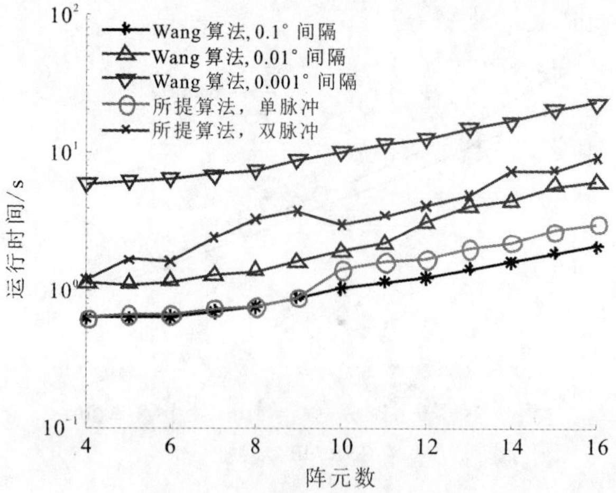

图 4.4　计算运行时间对比图

§4.2　速度场解相干的 MIMO 阵列角度估计

§4.2.1　引言

声矢量传感器[245-246]与传统的标量阵列信号参数估计相比，在相同孔径下具有更多的接收通道，从而提高了参数估计精度。声矢量传感器包括两个或三个正交共点的速度传感器和一个压力传感器。矢量传感器具有更多的输出信息，并且提供了更多的信号处理方式。本书结合速度矢量传感器和 MIMO 雷达提供的波形分集技术，给出了 MIMO 速度矢量阵列的信号模型[247-249]，并给出了该信号模型的 MUSIC[247-248]和 ESPRIT[249]角度估计方法。结果表明，这些附加自由度可以提高空间分辨率，增强参数估计能力。

尽管如前面所述许多学者已做出了杰出的贡献，但大多研究成果是在假设信号相互独立的条件下进行的。在实际应用中，由于多径信号来自同一目标的不同传播路径，相干信号的场景经常出现。对于相干源测向，可以使用空间平滑算法来恢复协方差矩阵的秩。空间平滑算法的缺点是会减小阵列孔径。因此，文献[250 - 252]提出了 VFS 算法来恢复协方差矩阵的秩，其算法不会减小阵列孔径。

本章将 VFS 算法扩展到 MIMO 雷达中，解决了速度矢量传感器 MIMO 雷达的相干源角度估计问题，并分析了去相关性能。去相关法的具体操作是将所有速度矢量内的同向传感器分成空间位置相同的三个子阵，然后对三个子阵的协方差矩阵进行平滑处理，恢复 SCM 的秩。

§4.2.2　速度矢量 MIMO 阵列参数估计信号模型

假设一个单基地 MIMO 雷达有 M 个发射标量传感器和 N 个三分量速度矢量接收传感器。假设发射阵列和接收阵列传感器放置在任意三维位置。N 个接收传感器中的每一个都是由三个相同但方向正交的速度矢量传感器组成的，分别沿着 x 轴，y 轴和 z 轴。将第 m 个传感器在发送阵列中的位置定义为 $\boldsymbol{l}_{tm}=[x_{tm}\ y_{tm}\ z_{tm}]$，将第 n 个传感器在接收阵列中的位置定义为 $\boldsymbol{l}_{rn}=[x_{rn}\ y_{rn}\ z_{rn}]$。发射传感器发送 M 个正交波形信号。在每个接收传感器中，回波与 M 个发射波形匹配滤波。假设 K 个不相关信号位于远场。经过匹配

滤波处理后，接收端的整个输出可以写成

$$\boldsymbol{x}(t)=\begin{bmatrix}\boldsymbol{x}_x(t)\\\boldsymbol{x}_y(t)\\\boldsymbol{x}_z(t)\end{bmatrix}=\begin{bmatrix}\boldsymbol{A\Phi}_x\\\boldsymbol{A\Phi}_y\\\boldsymbol{A\Phi}_z\end{bmatrix}\boldsymbol{s}(t)+\boldsymbol{n}(t) \tag{4.20}$$

$$\boldsymbol{A}=[\boldsymbol{\alpha}_t(\theta_1,\phi_1)\otimes\boldsymbol{\alpha}_r(\theta_1,\phi_1)\cdots\boldsymbol{\alpha}_t(\theta_K,\phi_K)\otimes\boldsymbol{\alpha}_K(\theta_K,\phi_K)] \tag{4.21}$$

$$\boldsymbol{\alpha}_t(\theta,\phi)=[\exp\{-j\kappa[\boldsymbol{l}_{t1}\boldsymbol{p}(\theta,\phi)]\}\cdots\exp\{-j\kappa[l_{tM}p(\theta,\phi)]\}]^{T} \tag{4.22}$$

$$\boldsymbol{\alpha}_r(\theta,\phi)=[\exp(-j\kappa[\boldsymbol{l}_{r1}\boldsymbol{p}(\theta,\phi)])\cdots\exp\{-j\kappa[\boldsymbol{l}_{rN}\boldsymbol{p}(\theta,\phi)]\}]^{T} \tag{4.23}$$

其中，式(4.22)和式(4.23)分别表示信息的发射导向矢量和接收导向矢量，$\boldsymbol{p}(\theta,\varphi)=[\sin\theta\cos\phi\ \sin\theta\sin\phi\ \cos\theta]^{T}$ 表示传播向量，θ 和 φ 分别代表 DOD 和 DOA，$\kappa=\frac{2\pi}{\lambda}$表示波数，$\boldsymbol{s}(t)=[s_1(t)\ s_2(t)\ \cdots\ s_K(t)]^{T}$ 代表相干信号源矢量，其中 $s_k(t)=\beta_k e^{j2\pi f_k t}$，$\beta_k$ 和 f_k 代表第 k 个信号的振幅和多普勒频率。假定噪声 $\boldsymbol{n}(t)$为白噪声，与均值为零且协方差为 $\sigma^2\boldsymbol{I}_{3MN}$ 的信号不相关。矩阵 $\boldsymbol{\Phi}_x$，$\boldsymbol{\Phi}_y$ 和 $\boldsymbol{\Phi}_z$ 等于：

$$\left.\begin{aligned}\boldsymbol{\Phi}_x&=\operatorname{diag}(\boldsymbol{p}_x)=\operatorname{diag}[\sin\theta_1\cos\phi_1\cdots\sin\theta_K\cos\phi_K]^{T}\\\boldsymbol{\Phi}_y&=\operatorname{diag}(\boldsymbol{p}_y)=\operatorname{diag}[\sin\theta_1\sin\phi_1\cdots\sin\theta_K\sin\phi_K]^{T}\\\boldsymbol{\Phi}_y&=\operatorname{diag}(\boldsymbol{p}_z)=\operatorname{diag}[\cos\theta_1\cdots\cos\theta_K]^{T}\end{aligned}\right\} \tag{4.24}$$

因为信源是相干的，所以直接利用式(4.20)的接收数据来计算 SCM 将是秩亏的。因此，无法直接应用传统的超分辨率算法。下面，笔者利用 VFS 算法来恢复 SCM 的秩。

§4.2.3 MIMO 阵列的“速度场平滑”算法

比较 x,y,z 轴上接收到的数据 $\boldsymbol{x}_x(t)$，$\boldsymbol{x}_y(t)$和 $\boldsymbol{x}_z(t)$发现，除了 $\boldsymbol{\Phi}_x$，$\boldsymbol{\Phi}_y$，$\boldsymbol{\Phi}_z$ 不同之外，其余部分具有相同的形式。因此，可以使用不同接收数据的加权平均来完成相干源的去相关，此方法称为 VFS 算法，因为在完成平滑处理时使用了速度场信息。实际上，平滑处理可以看作是空间平滑算法从空间域直接移植到速度场域的过程。下面用数学公式来描述 VFS 平滑处理。

首先分别计算接收到的 x,y,z 轴的数据的协方差矩阵，然后进行平滑处理，即

$$\boldsymbol{R}_{\text{smoothing}}=\frac{1}{3}\sum_{i=1}^{3}E\{\boldsymbol{x}_i(t)\boldsymbol{x}_i(t)^{\mathrm{H}}\}=\boldsymbol{A}\underbrace{\left(\frac{1}{3}\sum_{i=1}^{3}\boldsymbol{\Phi}_i\boldsymbol{R}_s\boldsymbol{\Phi}_i^{\mathrm{H}}\right)}_{\boldsymbol{R}_{\text{s\_smoothing}}}\boldsymbol{A}^{\mathrm{H}}+\sigma_n^2\boldsymbol{I}_{MN} \tag{4.25}$$

其中，$i=x,y,z$。把导向矩阵 $\boldsymbol{A}$ 设为满秩。在无噪声的情况下，矩阵 $\boldsymbol{R}_{\text{smoothing}}$ 的秩等于矩阵 $\boldsymbol{R}_{\text{s\_smoothing}}$ 的秩。信源 K 是相干的，所以可以设 $b_k(\tau)=g_k b(\tau)$，这里 g_k 表示非零复常数。则 SCM 等于 $\boldsymbol{R}_s=\sigma_s^2\boldsymbol{g}\boldsymbol{g}^{\mathrm{H}}$，其中 $\boldsymbol{g}=[g_1\ g_2\cdots g_K]^{\mathrm{T}}$。因此，$\boldsymbol{R}_{\text{s\_smoothing}}$ 也可表示为

$$\boldsymbol{R}_{\text{s\_smoothing}}=\frac{1}{3}\sum_{i=1}^{3}\boldsymbol{\Phi}_i\boldsymbol{R}_s\boldsymbol{\Phi}_i^{\mathrm{H}}=\frac{1}{3}\sigma_s^2\sum_{i=1}^{3}\boldsymbol{\Phi}_i\boldsymbol{g}\boldsymbol{g}^{\mathrm{H}}\boldsymbol{\Phi}_i^{\mathrm{H}}=\frac{1}{3}\sigma_s^2\boldsymbol{G}\left\{\sum_{i=1}^{3}\boldsymbol{p}_i\boldsymbol{p}_i^{\mathrm{H}}\right\}\boldsymbol{G}^{\mathrm{H}} \tag{4.26}$$

其中，$\boldsymbol{G}=\mathrm{diag}(\boldsymbol{g})$，注意矩阵 $\boldsymbol{G}$ 是对角矩阵，因此 $\mathrm{rank}(\boldsymbol{G})=K$。从文献[250][252]知道，$\boldsymbol{p}_i$ 和 $\boldsymbol{p}_j$ 对于任何 $\{(\theta_k,\varphi)_k,k=1\cdots K\}$ 线性独立，因此，$\mathrm{rank}\left(\sum_{i=1}^{3}\boldsymbol{p}_i\boldsymbol{p}_i^{\mathrm{H}}\right)=3$。通过以上分析，得出 $\mathrm{rank}(R_{\text{s\_smoothing}})=\min(K,3)$。此时，当信源数满足 $K\leqslant 3$ 时，SCM 的秩将会被恢复。因此，利用该算法可以解决三个相干信号入射到阵列的解相干问题。

本章所提算法需要注意两点：一是该算法仅能解三个相干源，如果使用前后项平滑算法，相干源的可分辨数将提高到六个；二是在上述过程中，没有使用阵列的结构信息，因此该方法适用于任意阵列结构。

§4.2.4　解相干性能分析

可以清楚地看到，去相关的性能取决于 $\boldsymbol{R}\overset{\text{def}}{=}\sum_{i=1}^{3}\boldsymbol{p}_i\boldsymbol{p}_i^{\mathrm{H}}$，我们称其为 AS-SCM。计算出 AS-SCM 的对角线元素，即 $\boldsymbol{R}_{i,i}=\sin^2\theta_i\cos^2\varphi_i+\sin^2\theta_i\sin^2\varphi_i+\cos^2\theta_i=1,i=1,\cdots,K$。AS-SCM 的非对角元素值等于：

$$\boldsymbol{R}_{i,j}=\sin\theta_i\cos\varphi_i\sin\theta_j\cos\varphi_j+\sin\theta_i\sin\varphi_i\sin\theta_j\sin\varphi_j+\cos\theta_i\cos\theta_j,i\neq j \tag{4.27}$$

从式(4.27)中可以看出 $\boldsymbol{R}_{i,j},i\neq j$，是二维 DOA 的函数。结果表明，所提出的 VFS 方法的去相关性能取决于二维 DOA。AS-SCM 的 $\boldsymbol{R}_{i,j},i\neq j$ 越小，解相关的性能会更好。如果 $\boldsymbol{R}_{i,j}=0,i\neq j$，相干信号将完全去相关。

§4.2.5 仿真结果分析

在本节中，使用仿真验证所提出算法的有效性。在下面的仿真中，将发射传感器设置为 $M=6$ 个，接收速度矢量传感器设置为 $N=6$ 个。

通过特征分解的特征值来验证 AS-SCM 去相关的有效性。假设有两个功率相同的相干源，信号源的角度分别为 $(\theta_1,\varphi_1)=(10°,15°)$ 和 $(\theta_2,\varphi_2)=(20°,25°)$。设置快拍数 $L=200$，SNR = 20 dB。为了验证所提出的算法适用于任意阵列结构，假设接收速度矢量传感器位于 x 轴，即：$[x_{r1},\cdots,x_{r6}]=\frac{1}{2}\lambda[2,3,7,10,11,12]$，$y_{rm}=0,z_{rm}=0,m=1,\cdots,6$，发射的标量传感器位于 y 轴，即：$[x_{t1},\cdots,x_{t6}]=\frac{1}{2}\lambda[0,1,3,6,9,13]$，$y_{tm}=0,z_{tm}=0,m=1,\cdots,6$。显然，阵列结构没有旋转不变性。因此，空间平滑算法是失效的。对于传统的比较，图 4.5 给出了具有相同数量的 AS-SCM 特征值的 SCM 的最大 MN 个特征值。从图 4.5 中可以清楚地看到，没有进行平滑处理的 SCM 只有一个大特征值。使用了 VFS 处理后，具有两个大特征值，它们的数目与相干源的数目相同。结果证明了所提出的 VFS 算法的有效性。

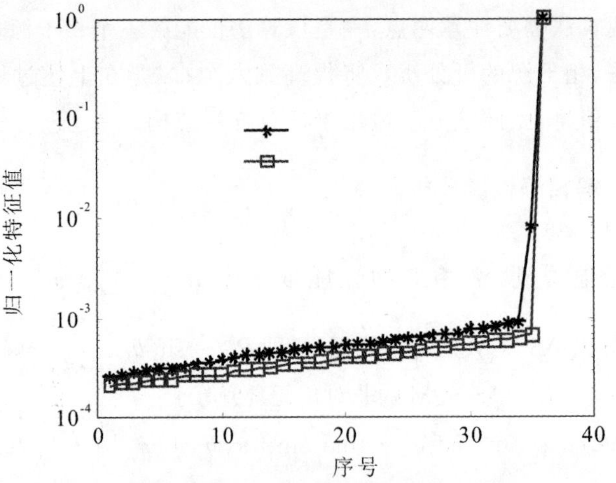

图 4.5 平滑处理和无平滑处理的协方差矩阵的特征值

在第二个仿真中，证明了这种平滑方法的性能取决于 DOA。图 4.6 显示了 AS-SCM 与 DOA 的非对角线元素之间的关系。为了不失一般性，设定$\theta=90°$，就有$R_{i,j}=\cos\varphi_i\cos\varphi_j+\sin\varphi_i\sin\varphi_j$。从图 4.6 中可以看出，非对角元素的值随 DOA 的变化而变化。

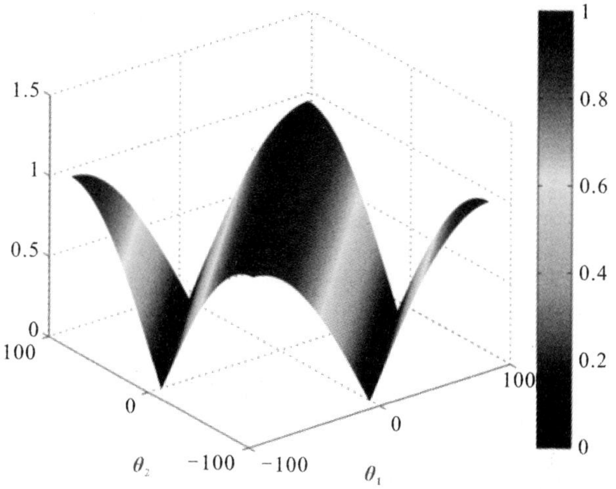

图 4.6　AS-SCM 与 DOA 的非对角线元素关系

在最后的仿真中，使用特征值分布和 DOA 估计的 RMSE 来进一步验证算法的有效性。方位角估计的均方根误差定义为 $\mathrm{RMSE}=\sqrt{\frac{1}{2\mathrm{Monte}}\sum_{q=1}^{\mathrm{Monte}}\sum_{k=1}^{K}(\hat{\varphi}_{k,q}-\varphi_{k,q})^2}$，$\hat{\varphi}_{k,q}$和$\varphi_{k,q}$分别表示第$q$次蒙特卡洛实验第$k$信源的方位角估计和实际方位角值。仰角估计的 RMSE 也使用类似的定义。这里设置两组相干信号源：$(\theta_1,\varphi_1,\theta_2,\varphi_2)=(10°,15°,20°,25°)$ 和 $(\theta_1,\varphi_1,\theta_2,\varphi_2)=(80°,90°,60°,0°)$。根据式(4.27)，第一组信号$R$的非对角元素等于$R_{12}=0.983\ 9$，第二组等于$R_{12}=0.086\ 8$。快拍数设置为$L=200$，蒙特卡洛实验等于 1 000。假设接收速度矢量传感器位于x轴，采用半波长均匀线阵，发射的标量传感器位于y轴，采用半波长均匀线阵。对于空间平滑算法，使用 ESPRIT 算法估计源的方位角和仰角。图 4.7 显示了两组信号的 AS-SCM 的特征值分布。图 4.8 显示了二维 DOA 估计的 RMSE 与 SNR 的关系曲线。由图 4.7 和图 4.8 可知，相关越性小，去相关的能力越强，估计精度也就越好。

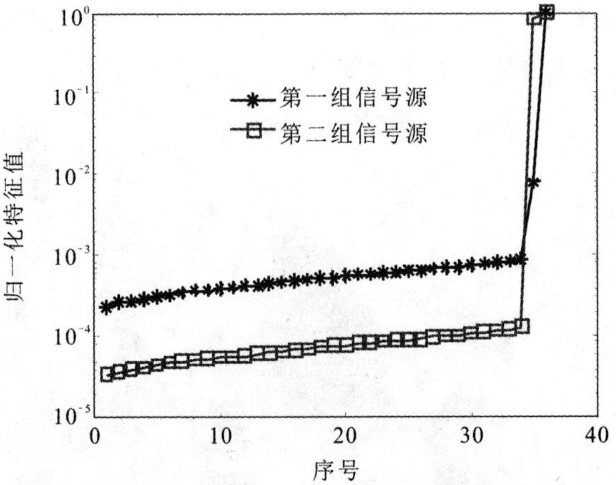

图 4.7　两组信号的 AS-SCM 的特征值分布

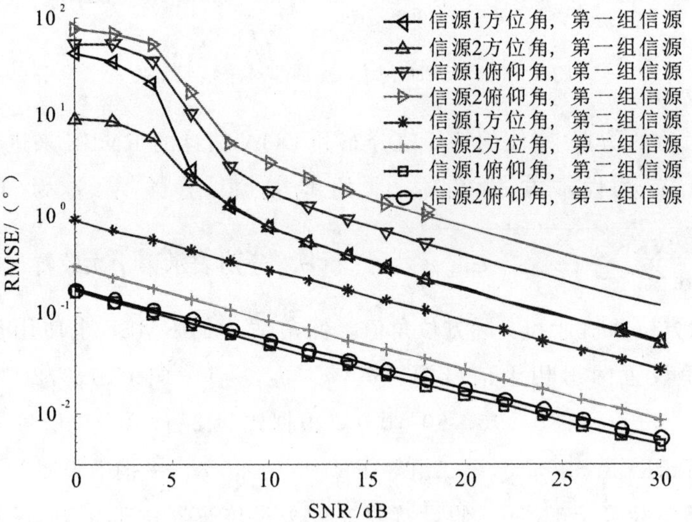

图 4.8　二维 DOA 估计的 RMSE 与 SNR 的关系曲线

§4.3　本 章 小 结

本章研究了 FDA-MIMO 雷达的距离和角度联合估计问题。在同样都是发射两个脉冲的前提下，所提的方法对比经典方法有三点好处：一是模糊距离变大；二是估计性能比经典算法略好；三是计算量更小，且所提方法能够得到角度和距离自动配对的结果。此外，本章还研究了接收速度矢量传感器对 MIMO 阵列的相干源 DOA 估计，提出的 VFS 预处理可以有效地恢复 AS-SCM 的秩。通过计算 AS-SCM 的非对角线元素来分析解相干性能。发现解相干性能与 DOA 直接相关，相关性越小估计精度就越好。此外，该算法适用于任意阵列结构。

第 5 章　基于波形分集的米波 MIMO 雷达测高方法研究

§ 5.1　基于 BOMP 预处理的米波 MIMO 雷达测高方法

§ 5.1.1　引言

米波雷达一般指工作频率在米波段的雷达，典型的有 VHF 波段雷达，米波雷达的发展由来已久，可追述到第二次世界大战时期，在防空预警中扮演着重要的角色。但米波雷达受限于波段太低、波束宽、波束掠地、地面反射回波幅度强，进而在低空目标探测中遇到相干多径、低信噪比等问题[253-254]，最终导致米波雷达仰角测量精度差，无法满足制导精度要求（众所周知，雷达目标高度实质是通过测量目标仰角后根据目标距离计算得到的）。因此，米波雷达的低仰角目标测高问题一直是阵列雷达信号处理中的重要问题之一。常规阵列的米波雷达测高目前有较为丰富的研究成果。其中有两类方法是研究的重点：一类是传统超分辨算法的移植和改进[255-256]，其中，地形因素是考虑重点[257]；另一类则是基于人工智能的应用[258]。

将 MIMO 体制应用到米波雷达中，研究米波 MIMO 雷达低仰角目标的 DOA 估计问题受到广泛关注。考虑平坦阵地光滑表面的镜面反射，不同于常规阵列雷达一个目标对应于两条接收路径，MIMO 雷达的一个目标对应于四条传输路径[146][259]，具体有：发射直达波-接收直达波；发射直达波-接收反射波；发射反射波-接收直达波；发射反射波-接收反射波。此时 MIMO 雷达不仅要面对相干源入射的问题，还要面临导向矢量相互渗透，或者有专家学者称之为锥角兼并的问题。故传统的相干源目标 DOA 估计或者双基地的 DOA 和 DOD 角度联合估计方法不能直接应用于 MIMO 阵列雷达的测高[259]，如以 MUSIC 为代表的超分辨算法技术在 MIMO 阵列雷达测高中无法直接应用。最大似然算法可直接处理相干信号，是米波 MIMO 雷达测角问题中常用的算法，但是涉及多维搜索的问题，使得本身就因 MIMO 阵列雷达自由度大而导致计算量大的问题更加严峻。鉴于此，文献[259]提出了一种降维的最大似然估计算法，该算法通过预先得到的目标距离和天线高度等先验信息，利用直达波信号与反射波信号之间存在的几何关系，进行直达波和反射波的合并，使得最大似然只需进行一维搜索便可完成对目标仰角的估计，极大地减少了运算量。

MIMO 雷达测高比较有效的方法有最大似然和广义 MUSIC 算法等，但是其涉及多维搜索的计算量较大。可以利用直达波和反射波具有的几何关系来降低搜索维数，但最大似然和广义 MUSIC 仍涉及整个空域的搜索，计算量仍然较大。本章提出基于 BOMP[260] 预处理的方法来降低计算量。首先对 MIMO 阵列接收数据稀疏化处理，将其变形至适合于 BOMP 算法的信号模型，然后利用粗栅格搜索得到角度粗估计，再以此为初始值中心，以 MIMO 雷达波束宽度作为搜索范围。这样处理的好处是既能保证最大似然和广义 MUSIC 高精度估计的优点，又能大大降低其计算量。

§5.1.2　信号模型

假设该 MIMO 雷达是一个收发共置的系统，其阵元有 M 个，且发射阵列为垂直均匀线阵，发射信号矢量为 $\boldsymbol{\phi}(t)\in\mathbb{C}^{M\times 1}$，MIMO 雷达区别于传统相控阵雷达，故设置发射信号相互正交，用数学表达式表示为 $\int_0^{T_p}\boldsymbol{\phi}(t)\phi(t)^{\mathrm{H}}\mathrm{d}t=\boldsymbol{I}_M$，其中积分符号中的 T_p 为雷达的脉冲持续时间，即脉冲宽度。对于连续波雷达则需要整个发射时间段的发射信号之间相互正交。假设 MIMO 雷达系统的低仰角反射区域是光滑平坦的反射面，如图 5.1 所示。

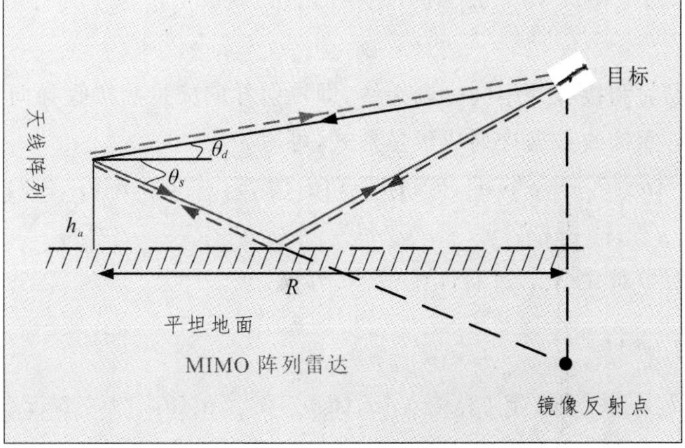

图 5.1　米波 MIMO 雷达低仰角测高镜面反射模型示意图

假设该系统的发射信号为水平极化信号，则其到达目标的信号可表示为

$$x(t)=[\boldsymbol{\alpha}_t(\theta_d)+e^{j\delta}\rho_h\boldsymbol{\alpha}_t(\theta_s)]^{T}\boldsymbol{\varphi}(t) \tag{5.1}$$

其中，θ_d 为直达波角度，θ_s 为反射波角度，$\boldsymbol{\alpha}_t(\theta)$为发射阵列导向矢量，即

$$\boldsymbol{\alpha}_t(\theta)=[1\ \exp[-j2\pi d\sin(\theta)/\lambda]\ \cdots\ \exp[-j2\pi(M-1)d\sin(\theta]/\lambda)]^{T} \tag{5.2}$$

其中，d 为阵元间隔，δ 为直达波和反射波波程差所引起的相位差，其表示为

$$\delta=\frac{4\pi h_a h_t}{R\lambda}$$

其中，h_a 为天线高度，h_t 为目标高度，λ 为入射波长，R 为目标垂直投影到地面的点到雷达的距离。需要注意的是波程差小于一个距离分辨单元，故在距离上区分不了直达和反射波。ρ_h 为水平极化波的反射系数，其值等于：

$$\rho_h=\frac{\sin\theta_d-\sqrt{\varepsilon-\cos^2\theta_d}}{\sin\theta_d+\sqrt{\varepsilon-\cos^2\theta_d}} \tag{5.3}$$

其中，ε 为表面复介电常数，通常可由相对介电常数 ε_r 和表面物质传导率 σ_e 表示：

$$\varepsilon=\varepsilon_r-j\,60\lambda\sigma_e$$

另外，$\beta(\tau)=\alpha\exp(j2\pi f_d\tau)$表示不同脉冲之间的未知复反射系数，假设服从 Swelling 2 型分布。则第 m 个接收阵元的接收数据为

$$\begin{aligned}z_m(t,\tau)&=[a_{r,m}(\theta_d)+e^{j\delta}a_{r,m}(\theta_s)]\beta(\tau)x(t)+n_m(t,\tau)\\&=[a_{r,m}(\theta_d)+e^{j\delta}\rho_h a_{r,m}(\theta_s)]\beta(\tau)[a_t(\theta_d)+e^{j\delta}\rho_h\,a_t(\theta_s)]^{T}\varphi(t)+n_m(t,\tau)\end{aligned} \tag{5.4}$$

假设发射和接收共用同一副天线，即发射导向矢量和接收导向矢量相同。将所有的阵元接收数据排列成矢量形式，可写为

$$\begin{aligned}\boldsymbol{z}(t,\tau)=&[\boldsymbol{\alpha}_r(\theta_d)+e^{j\delta}\rho_h\boldsymbol{\alpha}_r(\theta_s)]\beta(\tau)\,[\boldsymbol{\alpha}_t(\phi,\theta_d)+e^{j\delta}\rho_h\boldsymbol{\alpha}_t(\phi,\theta_s)]^{T}\boldsymbol{\varphi}(t)+\\&\boldsymbol{n}(t,\tau)\in\mathbb{C}^{M\times 1}\end{aligned} \tag{5.5}$$

利用发射信号对式(5.5)进行匹配滤波，可得

$$\begin{aligned}\boldsymbol{Z}&=\int z(t,\tau)\varphi\,(t)^{H}\\&=[\boldsymbol{\alpha}_r(\theta_d)+e^{j\delta}\rho_h\boldsymbol{\alpha}_r(\theta_s)]\,\beta(\tau)\,[\boldsymbol{\alpha}_t(\theta_d)+e^{j\delta}\rho_h\boldsymbol{\alpha}_t(\theta_s)]^{T}+\boldsymbol{N}(\tau)\in\mathbb{C}^{M\times M}\end{aligned} \tag{5.6}$$

对上述数据进行矢量化操作可得

$$\mathrm{vec}(\boldsymbol{Z})=[\boldsymbol{\alpha}_t(\theta_d)+\mathrm{e}^{\mathrm{j}\delta}\rho_h\boldsymbol{\alpha}_t(\theta_s)]\otimes[\boldsymbol{\alpha}_r(\theta_d)+\mathrm{e}^{\mathrm{j}\delta}\rho_h\boldsymbol{\alpha}_r(\theta_s)]\beta(\tau)+\mathrm{vec}[\boldsymbol{N}(\tau)]\in\mathbb{C}^{M^2\times 1} \tag{5.7}$$

首先假设原始噪声是零均值高斯随机过程，则有

$$E[\boldsymbol{n}(t_1,\tau)\boldsymbol{n}(t_2,\tau)^{\mathrm{H}}]=\sigma^2\boldsymbol{I}_M\omega(t_1-t_2)$$

其中，ω 为冲激函数。则匹配滤波后噪声等于：

$$\boldsymbol{N}(\tau)=\int_0^{T_p}\boldsymbol{n}(t,\tau)\boldsymbol{\varphi}(t)^{\mathrm{H}}\mathrm{d}t \tag{5.8}$$

根据文献[146]的结论可得，经过匹配滤波、矢量化操作之后，噪声仍然是白噪声。

§5.1.3　基于 BOMP 预处理的米波 MIMO 雷达测高

研究表明常规相控阵列中的“相干信号”与“多径信号”的关系基本可以等同，所有对相干信号处理有效的算法均可应用于多径信号，如典型的空间平滑解相干算法。但 MIMO 雷达中“相干信号”与“多径信号”的关系不等效，因为 MIMO 雷达存在信号相互渗透的现象，这导致传统空间平滑无法利用，进而导致传统基于子空间的经典超分辨算法 ESPRIT 和 MUSIC 无法直接利用。因此，无需解相干类的超分辨算法，如最大似然算法和广义 MUSIC 算法，在 MIMO 雷达测高中得到了充分应用，是两种行之有效的测高算法。下面简要回顾这两种算法。

1.最大似然

最大似然估计是阵列信号处理中最常用和最有效的参数估计方法之一。从其参数选取准则中可以看出，其算法并不受信号间相关性的影响。具体推导详见参考文献[261]的第 148 页式(5.2.10)。最大似然估计准则可表示为

$$\hat{\boldsymbol{\theta}}=-\underset{\boldsymbol{\theta}}{\mathrm{argmax}}\mathrm{Tr}[\boldsymbol{P}_{A(\theta)}\hat{\boldsymbol{R}}] \tag{5.9}$$

其中，$\hat{\boldsymbol{\theta}}$ 为 $\boldsymbol{\theta}$ 的最大似然估计，$\boldsymbol{A}(\boldsymbol{\theta})$为导向矢量矩阵，$\hat{\boldsymbol{R}}$ 为阵列输出信号的协方差矩阵估计值，$\boldsymbol{P}_{\boldsymbol{A}(\boldsymbol{\theta})}$ 为投影到导向矢量矩阵 $\boldsymbol{A}(\boldsymbol{\theta})$的列向量张成空间的投影矩阵，其值等于：

$$\boldsymbol{P}_{\boldsymbol{A}(\theta)}=\boldsymbol{A}(\boldsymbol{\theta})[\boldsymbol{A}^{\mathrm{H}}(\boldsymbol{\theta})\boldsymbol{A}(\boldsymbol{\theta})]^{-1}\boldsymbol{A}^{\mathrm{H}}(\boldsymbol{\theta}) \tag{5.10}$$

将式(5.7)中 MIMO 雷达接收数据中的信号源和导向矢量重新组合成如下形式：

$$[\boldsymbol{\alpha}_t(\theta_d)+e^{j\delta}\rho_h\boldsymbol{\alpha}_t(\theta_s)]\otimes\{\boldsymbol{\alpha}_r(\theta_d)+e^{j\delta}\rho_h\boldsymbol{\alpha}_r(\theta_s)\}\beta(\tau)=$$

$$\begin{bmatrix}\boldsymbol{\alpha}_t(\theta_d)\otimes\boldsymbol{\alpha}_r(\theta_d) & \boldsymbol{\alpha}_t(\theta_d)\otimes\boldsymbol{\alpha}_r(\theta_s)\\ \boldsymbol{\alpha}_t(\theta_s)\otimes\boldsymbol{\alpha}_r(\theta_d) & \boldsymbol{\alpha}_t(\theta_s)\otimes\boldsymbol{\alpha}_r(\theta_s)\end{bmatrix}^{T}\begin{bmatrix}1\\ e^{j\delta}\rho_h\\ e^{j\delta}\rho_h\\ e^{j2\delta}\rho_h{}^2\end{bmatrix}\beta(\tau) \tag{5.11}$$

则可将式(5.7)中的接收数据写成如下形式：

$$\boldsymbol{r}=\boldsymbol{Ay}+\boldsymbol{\varepsilon} \tag{5.12}$$

其中，$\boldsymbol{r}=vec(\boldsymbol{Z})$，$\boldsymbol{A}=\begin{bmatrix}\boldsymbol{\alpha}_t(\theta_d)\otimes\boldsymbol{\alpha}_r(\theta_d) & \boldsymbol{\alpha}_t(\theta_d)\otimes\boldsymbol{\alpha}_r(\theta_s)\\ \boldsymbol{\alpha}_t(\theta_s)\otimes\boldsymbol{\alpha}_r(\theta_d) & \boldsymbol{\alpha}_t(\theta_s)\otimes\boldsymbol{\alpha}_r(\theta_s)\end{bmatrix}^{T}$，$\boldsymbol{\varepsilon}=\mathrm{vec}[\boldsymbol{N}(\tau)]\in\mathbb{C}^{M^2\times 1}$，$\boldsymbol{y}=[1\ e^{j\delta}\rho_h\ e^{j\delta}\rho_h\ e^{j2\delta}\rho_h{}^2]^{T}\beta(\tau)$。通过上述定义，可以看出对应于文书的 MIMO 阵列雷达，式(5.9)中最大似然 $\boldsymbol{\theta}=[\theta_d\ \theta_s]$ 为信源的直达波和反射波。$\mathbf{A}(\boldsymbol{\theta})=\mathbf{A}$，$\hat{\mathbf{R}}=E\{\boldsymbol{rr}^{H}\}$。

式(5.10)中的最大似然估计求解过程涉及二维非线性搜索，其计算量很大。可采用直达波和反射波的数学关系式来降低搜索维度，其关系式为：

$$\theta_s=-\arcsin(\sin\theta_d+2h_a/R) \tag{5.13}$$

2.广义 MUSIC

传统 MUSIC 算法是建立在信号和噪声子空间的正交基础之上。但在 MIMO 雷达中，受相干信号的影响，导致导向矢量与噪声子空间并不正交。即使采用某种方法去相干，其中导向矢量信号相互渗透，即 $\boldsymbol{a}_t(\theta_d)\otimes\boldsymbol{a}_r(\theta_s)$，从而导致导向矢量与噪声子空间并不正交，故传统 MUSIC 算法无法直接应用。广义 MUSIC 亦建立在信号子空间和噪声子空间正交基础上，但并没有利用导向矢量而是利用导向矩阵与噪声子空间正交的原理。需要注意"导向矢量""导向矩阵""噪声子空间"之间的相互关系，见表 5.1。

表 5.1　"导向矢量""导向矩阵""噪声子空间"之间的相互关系

噪声子空间	导向矢量	导向矩阵(信号子空间)
常规阵列雷达	正交(MUSIC)	正交(广义 MUSIC)
MIMO 阵列雷达	不正交(MUSIC)	正交(广义 MUSIC)

从表 5.1 中可清晰看出，对于 MIMO 阵列雷达，传统 MUSIC 不可用，而广义 MUSIC 可用。第二章的推导，根据接收数据的特征值分解得到噪声子空间 E_n，根据文献[261]的推导，广义 MUSIC 谱为

$$P(\boldsymbol{\theta})=\frac{\det[\boldsymbol{A}^{\mathrm{H}}(\boldsymbol{\theta})\boldsymbol{A}(\boldsymbol{\theta})]}{\det[\boldsymbol{A}^{\mathrm{H}}(\boldsymbol{\theta})\boldsymbol{E}_n\boldsymbol{E}_n^{\mathrm{H}}\boldsymbol{A}(\boldsymbol{\theta})]} \tag{5.14}$$

式(5.14)同样涉及二维搜索，可采用直达波和反射波关系得到一维搜索谱估计。

3.基于 BOMP 的米波 MIMO 雷达测高粗估计

式(5.9)和式(5.14)的最大似然和广义 MUSIC 算法需全空域搜索，搜索范围太大。下面介绍一种单快拍条件下即可得到仰角粗估计的方法来缩小搜索范围，以此减少计算量。首先令式(5.12)中的 $\tau=1$，即单快拍。对导向矢量和接收信号源采用稀疏表示。构造稀疏过完备矩阵 $\boldsymbol{A}(\boldsymbol{\psi})$，$\boldsymbol{\psi}=(\theta_d,\theta_s)$，如此构造过完备矩阵，则需要构造二维矩阵，其计算量很大，这里同样可利用直达波和反射波之间的关系式(5.13)来降低过完备字典的维数，即 $\boldsymbol{\psi}=(\theta_d)$，$L$ 表示字典的长度，其中 L 远大于目标个数，这里设目标个数为 K（事实上对于测高模型，目标为 1）。则式(5.12)可以写成

$$\boldsymbol{r}=\boldsymbol{A}(\boldsymbol{\psi})\boldsymbol{z}+\boldsymbol{\varepsilon} \tag{5.15}$$

其中，$\boldsymbol{z}$ 表示 K 块稀疏矢量。其中块的长度等于 4。式(5.15)可利用 BOMP 算法去寻找支撑位置，从而计算出角度。下面简单给出 BOMP 算法与式(5.15)的对应关系。

将 $(\theta_{d,k})_{k=1}^{K}$ 划分成 L 个格子，其中 $L\gg K$。这样 $(\theta_{d,l})_{l=1}^{L}$ 对应的块稀疏矢量 $\boldsymbol{z}\triangleq[(\boldsymbol{z}^{[1]})^{\mathrm{T}}\cdots(\boldsymbol{z}^{[l]})^{\mathrm{T}}\cdots(\boldsymbol{z}^{[L]})^{\mathrm{T}}]^{\mathrm{T}}\in\mathbb{C}^{4L\times 1}$，其中 $\boldsymbol{z}^{[l]}=[z_1^{[l]}\ z_2^{[l]}\ z_3^{[l]}\ z_4^{[l]}]^{\mathrm{T}}\in\mathbb{C}^{4\times 1}$。这里假设目标落在搜索网格上（至于目标落在两个格子中间的情况，由于是粗估计，并不影响最终结果），则有

$$\forall\, 1\leqslant k\leqslant K,\ \exists\, 1\leqslant l\leqslant L,\ s.t.\boldsymbol{z}^{[l]}=\boldsymbol{y}^{[k]} \tag{5.16}$$

所以对于 $\forall\, 1\leqslant l\leqslant L$，$\boldsymbol{z}^{[l]}$ 只有两种取值：

$$\left.\begin{array}{l} z_i^{[l]}=0\ (1\leqslant i\leqslant 4)\\ z_i^{[l]}\neq 0\ (1\leqslant i\leqslant 4)\end{array}\right\} \tag{5.17}$$

目标则转化为找出块稀疏矢量 $\boldsymbol{z}$ 中目标的 K 个支撑位置，支撑位置就是目标的稀疏字典中仰角的位置。现将优化问题转化为

$$\left.\begin{array}{l} \underset{\boldsymbol{z}}{\operatorname{argmin}}\ \|\boldsymbol{z}\|_{B,0}\\ \mathrm{s.t.}\boldsymbol{A}_a(\boldsymbol{\psi})\boldsymbol{z}=\boldsymbol{r}-\hat{\boldsymbol{\varepsilon}}\end{array}\right\} \tag{5.18}$$

其中，$\|\boldsymbol{z}\|_{B,0}\triangleq\#\{l\,|\,\boldsymbol{z}[l]\neq 0,1\leqslant l\leqslant L\}$，$\hat{\boldsymbol{\varepsilon}}$ 是 $\boldsymbol{\varepsilon}$ 的估计值，可用协方差矩阵分解得到其噪声的估计值。式(5.18)的优化目标函数是为寻找一个最稀疏的解（故用 l_0 范数）。式(5.18)的约束条件是要求这个稀疏解满足观测条件（即

与采集数据吻合)。当接收数据的快拍数为 1 时,上式就是一个标准的 K 块稀疏问题,其支撑位置可用 BOMP 算法求出,结果记为$\hat{z}[k]$,$k=1\cdots K$,然后容易计算出目标的位置 θ_k^{coarse}。为了便于说明,这里给出支撑位置的概念,指的是目标在稀疏字典中的非零位置,在本书中即是目标的位置,即

$$\hat{\theta}_k^{\text{coarse}}=V(\hat{z}[k]-1)/4,\quad k=1\cdots K \tag{5.19}$$

其中,V 表示构造稀疏字典中的间隔度数。当快拍数多时,可用上述方法取得的结果求平均,事实上这里的结果仅作为粗估计的结果,为减少计算量可直接采用单快拍 BOMP 算法。

4.米波 MIMO 雷达测高的精确估计

此时,米波 MIMO 雷达的仰角初始估计值已经得到。在最大似然和广义 MUSIC 算法中以此为中心来确定搜索范围。下面给出搜索范围的基本原则。首先定义搜索范围为 $\theta=(\theta^{\text{coarse}}+\theta_1,\theta^{\text{coarse}}-\theta_1)$,核心是确定 θ_1 的取值。这里的准则定为半波束宽度,此时搜索范围等于一个波束宽度。如果目标功率降到波束宽度之外是检测不到目标的,故在单个波束宽度内搜索是合理的。MIMO 雷达的波束宽度为:$\theta_{mb}=\dfrac{50.7\lambda}{Nd\cos\theta_d}$,$\theta_1=\dfrac{\theta_{mb}}{2}$,$N$ 表示 MIMO 雷达的等效孔径阵元数,对于收发共置的 MIMO 雷达,有效孔径阵元数 $N=2M-1$,对于非等距线阵亦容易计算得到,这里不再赘述。测高的步骤如下:

第一步:利用 BOMP 得到初始估计,并确定搜索范围。表 5.2 给了出 BOMP 算法的计算流程。

表 5.2　BOMP 算法的计算流程

步　骤	执行内容
输　入	匹配滤波后的矢量化数据、角度网格数、目标数
初始化	用接收数据初始化残差、用角度网格数初始化字典、初始化支撑集
迭　代	1.利用残差和字典计算投影; 2.根据投影寻找块最大的坐标值,并将此值坐标放入块支撑集; 3.利用块支撑集更新残差; 4.迭代步骤 1 至步骤 3,迭代次数达到目标数停止
输　出	利用块支撑集计算块支撑向量

第二步：利用第一步搜索范围，使用最大似然和广义 MUSIC，得到最终角度估计值，然后根据目标距离转化成目标高度值。

本算法需要注意的有：对于米波 MIMO 阵列雷达模型，当目标的仰角小于波束宽度的二分之一时，Rank($\boldsymbol{A}$)=4，Rank($\boldsymbol{y}$)=1。如果从常规阵列信号处理的角度考虑，可认为是 4 个目标，且相互相干。但是仔细观察会发现导向矩阵中存在信号相互渗透的现象，实际上只有两个不一样的角度入射到阵列，从参数估计的角度来考虑，可认为是两个信号源，故上面所提到的最大似然算法和广义 MUSIC 算法，均涉及二维搜索。同样是因为相互渗透的原因导致信号的空间平移性不存在，进而导致空间平滑解相干算法不可用。既然不具备空间平移不变性，则传统超分辨算法 ESPRIT 亦不可用。对于经典超分辨算法 MUSIC 亦不可直接应用，后续会分析无需解相干的经典超分辨算法 MUSIC 在米波 MIMO 雷达测高中的应用。另一方面，本章算法是建立在平坦阵地模型之上的。对于粗糙反射表面、地形起伏等复杂阵地条件下的米波 MIMO 雷达测高，本章算法不能直接应用。关于米波 MIMO 雷达复杂阵地条件下的研究，已有的成果有地面反射系数未知的交替搜索 MUSIC 算法、反射面高度未知和多径数未知的条件下利用缩放字典逐层逼近的方法，以及基于秩 1 约束与压缩感知的低角目标测高算法等，这些算法能够解决部分复杂场景下的米波雷达测高问题。后续作者将在此基础上，研究不同阵元对应不同反射点高度，即同一阵列雷达反射点起伏的测高估计问题，这也是测高最为复杂的问题之一。

§5.1.4　仿真结果分析

仿真一：BOMP 预处理的支撑位置恢复情况。

考虑该米波 MIMO 阵列雷达的发射阵元数 $M=10$，阵元间距为半波长布局。入射频率为 300 MHz，入射波长 $\lambda=1$ m，目标直达波角度为 5°，反射角角度根据公式(5.13)计算得到。信噪比 SNR=0 dB，其中天线高度 $h_a=5$ m，目标高度 $h_t=7\ 000$m，设置淡水场景，则可设置反射系数中的介电常数 $\varepsilon_r=80$ 和表面物质传导率 $\sigma_e=0.2$。其中[：n°：]表示每间隔 n°取一个数值来构造稀疏字典，这里设 $n=1$，则目标的理论仰角值的第一个支撑位置为 5×4 +1=21。图 5.2 给出了其中一次独立实验的支撑位置估计结果。可看出该 BOMP 算法能够正确估计出目标仰角值。

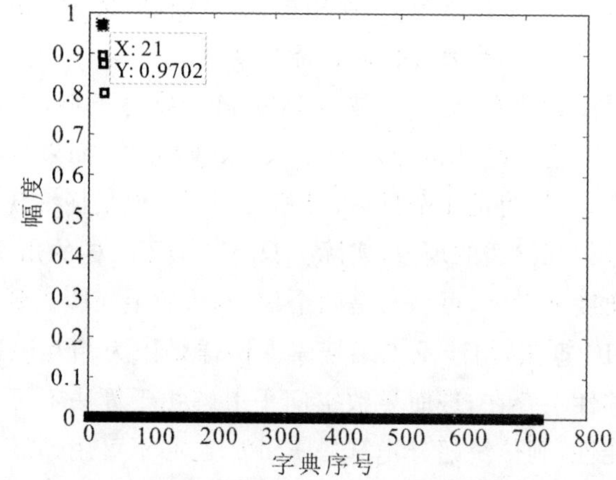

图 5.2 基于 BOMP 稀疏恢复的支撑位置估计结果

仿真二:搜索范围选择的正确性验证。

考虑该米波 MIMO 雷达的发射阵元数 $M=10$,入射频率为 300 MHz,入射波长 $\lambda=1$ m,阵元间距为半波长布局。目标直达波角度为 5°,反射角角度根据公式(5.13)计算得到。信噪比 SNR=5 dB,快拍数为 10 个,天线高度 $h_a=5$ m,目标高度 $h_t=7\ 000$ m,设置淡水场景,则可设置反射系数中的介电常数 $\varepsilon_r=80$ 和表面物质传导率 $\sigma_e=0.2$。图 5.3 给出了最大似然和广义 MUSIC 的全空域搜索的一次估计结果。

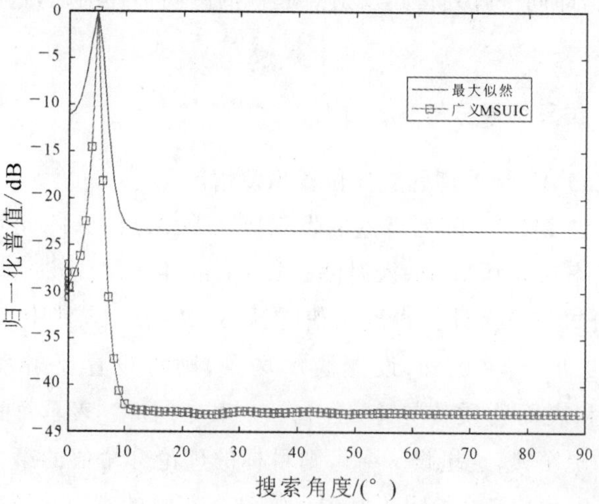

图 5.3 最大似然和广义 MUSIC 的全空域搜索的一次估计结果

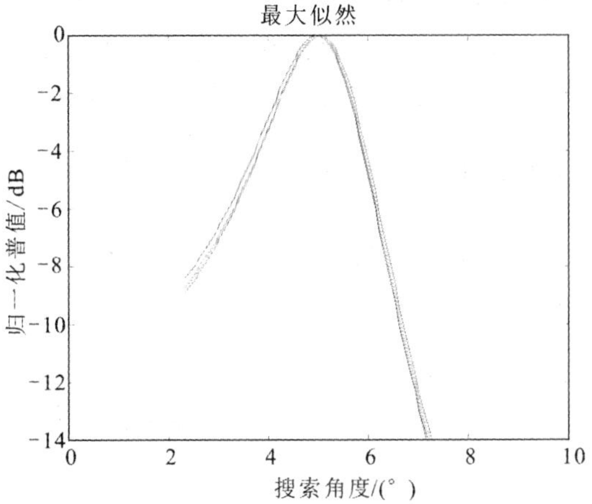

图 5.4　10 次独立实验的基于 BOMP 预处理后最大似然的搜索范围结果

图 5.4 给出 10 次独立实验的基于 BOMP 预处理后最大似然的搜索范围结果。图 5.5 给出 10 次独立实验的基于 BOMP 预处理后广义 MUSIC 的搜索范围结果。从图中可以看出本书的算法能够正确地缩小搜索范围，从而减少计算量。

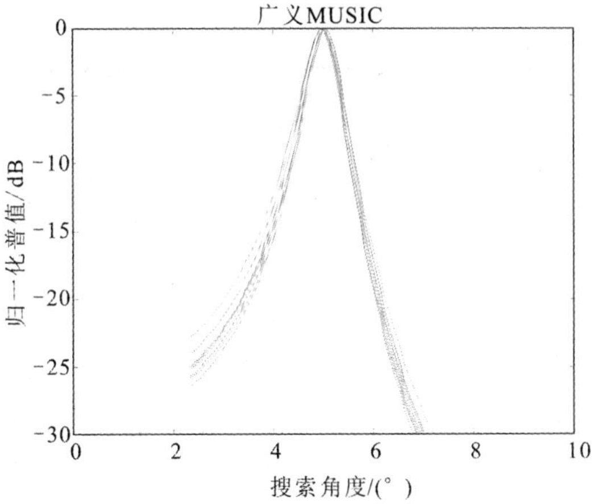

图 5.5　10 次独立实验的基于 BOMP 预处理后广义 MUSIC 的搜索范围结果

仿真三：最大似然和广义 MUSIC 算法仰角与高度估计结果。

考虑该米波 MIMO 雷达的发射阵元数 $M=10$，入射频率为 300 MHz，入射波长 $\lambda=1$ m。阵元间距为半波长布局。目标直达波角度为 5°，反射角角度根据公式(5.13)计算得到。信噪比变化,快拍数为 10 个,天线高度 $h_a=5$ m,目标高度 $h_t=7\ 000$ m,设置淡水场景,则可设置反射系数中的介电常数 $\varepsilon_r=80$ 和表面物质传导率 $\sigma_e=0.2$。完成 1 000 次蒙特卡洛实验,并定义衡量估计性能的均方根误差(RMSE)为:$\mathrm{RMSE}=\sqrt{\frac{1}{\mathrm{Monte}}\sum_{p=1}^{\mathrm{Monte}}E[(\hat{\alpha}_p-\alpha_p)^2]}$,其中,$\hat{\alpha}=\hat{\theta}_d$ 为直达波角度估计值,$\alpha=\theta_d$ 为目标角度真实值。$\hat{\alpha}=\hat{h}_t$ 为目标高度估计值,$\alpha=h_t$ 为目标高度真实值。

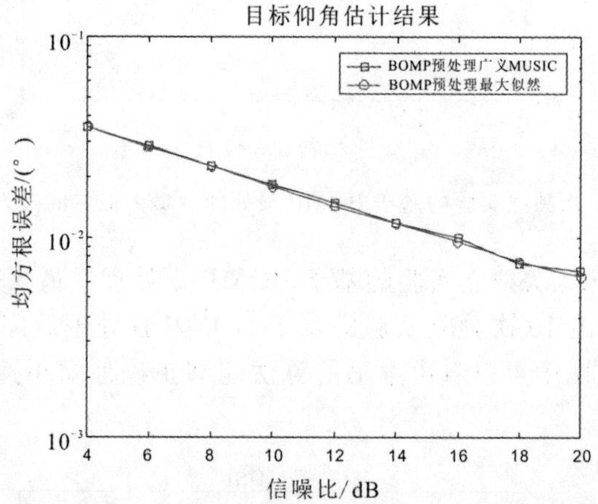

图 5.6　最大似然和广义 MUSIC 算法的角度估计 RMSE 随 SNP 变化曲线

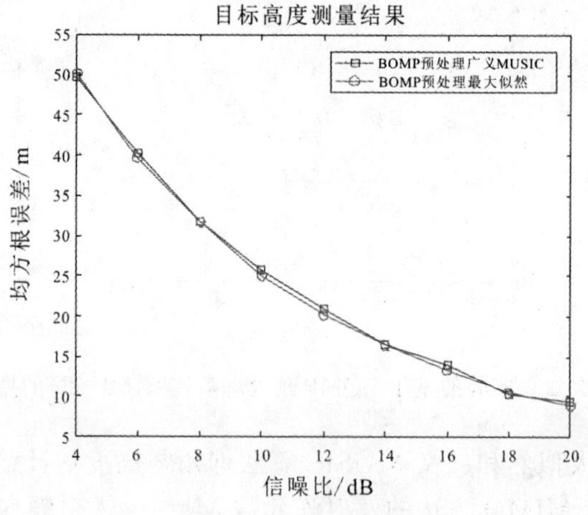

图 5.7　最大似然和广义 MUSIC 算法的高度测量 RMSE 随 SNR 变化曲线

图 5.6 和图 5.7 中给出基于 BOMP 预处理后的最大似然和广义 MUSIC 算法的角度估计的 RMSE 随 SNR 变化曲线和高度测量的 RMSE 随 SNR 变化曲线，从图中可以看出这两种算法能够正确应用在 MIMO 雷达对目标的测高领域。

仿真四：预处理和非预处理算法运算时间的比较。

考虑该米波 MIMO 雷达的发射阵元数从 10 变化到 20，入射频率为 300 MHz，入射波长 $\lambda=1$ m，阵元间距为半波长布局。目标直达波角度为 5°，反射角角度根据公式(5.13)计算得到。信噪比 SNR＝5 dB，快拍数为 10 个，天线高度 $h_a=5$ m，目标高度 $h_t=7\ 000$ m，设置淡水场景，则可设置反射系数中的介电常数 $\varepsilon_r=80$ 和表面物质传导率 $\sigma_e=0.2$。图 5.8 所示为最大似然和广义 MUSIC 的全空域搜索的运算时间和基于 BOMP 预处理后最大似然和广义 MUSIC 的运算时间对比。从图 5.8 中可以看出，本书算法能够有效减少计算量，阵元数越多，本书算法的优势越明显。

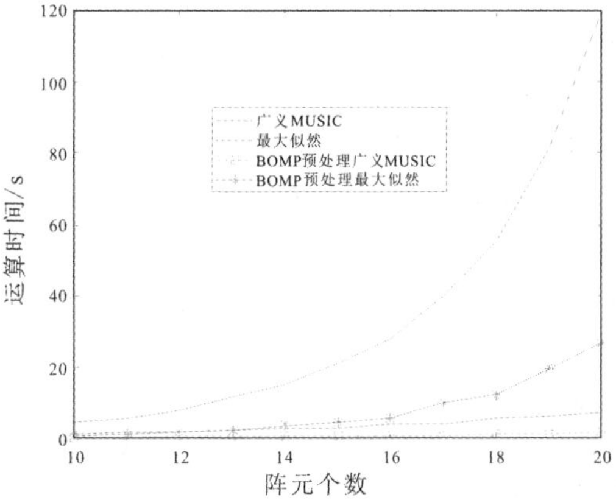

图 5.8　预处理和非预处理算法运算时间比较结果

§5.2　复杂阵地精确信号模型的米波 MIMO 雷达测高方法

§5.2.1　引言

对于米波 MIMO 阵列雷达，测高方法包括最大似然法[146]、广义 MUSIC 法、智能算法[158]、波瓣分裂算法[262]、时间反转等[263]。这些方法可以解决米波 MIMO 雷达测高问题。另外，由于 MIMO 阵列雷达的自由度高，计算量大，文献[18－20]提出了波束空间、矩阵波束和降维根降维等方法来减少计算量，取得了良好的测高效果。然而，已有算法中复杂地形上的反射点只定位在一个点上，所有元素的反射系数都是相同的，这与实际信号模型严重不符。

在低仰角测高中，一个合适的信号模型是非常重要的。为了与实际情况相匹配，本章提出了一种新的复杂地形信号模型，该信号模型考虑了每个阵元、每个目标的反射点。在该信号模型中，每个阵元的反射点均不相同，其反射系数因反射点不同也不相同。首先推导了平坦地面下考虑每个阵元反射点的情况，然后对传统的起伏阵地测高模型进行了简要回顾，最后推导了考虑每个阵元、每个信号反射点的起伏阵地测高模型。经理论推导，该信号模型较贴近于实际信号模型。针对精确的地形信号模型，推导了最大似然法和广义 MUSIC 法在 MIMO 雷达中的应用。

§5.2.2 复杂地形 MIMO 雷达测高精确模型建立

1.平坦阵地的测高模型

假设一个阵列雷达各向同性辐射，最高的阵元设置为参考阵元，设置目标信号为 $s(t)$，假设目标无衰减，只有延迟引起的相位变化。假设接收阵元的增益相同，特性也相同，通道一致性良好。经典平坦阵地的信号接收示意图如图 5.9 所示。

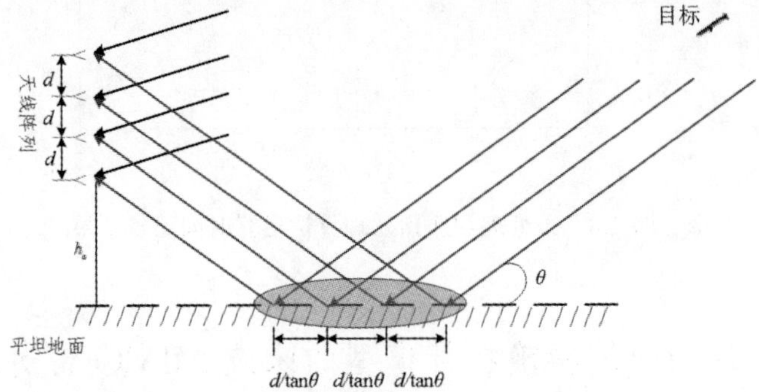

图 5.9 经典平坦阵地的信号接收示意图

参考阵元接收到的直达波信号为

$$s(t)e^{-j2\pi f\frac{R_{d,1}}{c}},R_{d,1}=\sqrt{(h_t-h_1)^2+R^2}\approx\bar{R}-h_1\sin\theta_d \tag{5.20}$$

其中，$\bar{R}$ 为目标斜距，天线高度 $h_1=h_a+(N-1)d$，h_a 为天线最低的天线阵元高度，d 为阵元间等间距的距离，N 为阵列的阵元总数，R 为天线阵列与目标在水平面上的投影的水平距离，$R_{d,1}$ 为目标到参考阵元的波程，h_t 为目标到水平面的距离。θ_d 为直达波的目标角度。

从上往下数第二个阵元的直达波信号为

$$s(t)\mathrm{e}^{-\mathrm{j}2\pi f\frac{Rd,2}{c}},R_{d,2}=\sqrt{(h_t-h_2)^2+R^2}\approx\bar{R}-h_2\sin\theta_d \tag{5.21}$$

第 N 个阵元的直达波信号为

$$s(t)\mathrm{e}^{-\mathrm{j}2\pi f\frac{Rd,N}{c}},R_{d,N}=\sqrt{(h_t-h_N)^2+R^2}\approx\bar{R}-h_N\sin\theta_d \tag{5.22}$$

所以直达波信号矢量可以写成：

$$\begin{aligned}\boldsymbol{x}_d(t)&=[s(t)\mathrm{e}^{-\mathrm{j}2\pi f\frac{Rd,1}{c}}\ s(t)\mathrm{e}^{-\mathrm{j}2\pi f\frac{Rd,2}{c}}\cdots s(t)\mathrm{e}^{-\mathrm{j}2\pi f\frac{Rd,N}{c}}]^{\mathrm{T}}\\&=s(t)\mathrm{e}^{-\mathrm{j}2\pi f\frac{\bar{R}}{c}}[\mathrm{e}^{\mathrm{j}2\pi f\frac{h1\sin\theta d}{c}}\ \mathrm{e}^{\mathrm{j}2\pi f\frac{h2\sin\theta d}{c}}\cdots \mathrm{e}^{-\mathrm{j}2\pi f\frac{hN\sin\theta d}{c}}]^{\mathrm{T}}\\&=s(t)\mathrm{e}^{-\mathrm{j}2\pi f\frac{\bar{R}-(ha+(N-1)d)\sin\theta d}{c}}[1\ \mathrm{e}^{-\mathrm{j}2\pi f\frac{d\sin\theta d}{c}}\cdots \mathrm{e}^{-\mathrm{j}2\pi f\frac{(N-1)d\sin\theta d}{c}}]^{\mathrm{T}}\\&=s(t)\mathrm{e}^{-\mathrm{j}2\pi f\frac{Rd,1}{c}}[1\ \mathrm{e}^{-\mathrm{j}2\pi f\frac{d\sin\theta d}{c}}\cdots \mathrm{e}^{-\mathrm{j}2\pi f\frac{(N-1)d\sin\theta d}{c}}]^{\mathrm{T}}\\&=s(t)\mathrm{e}^{-\mathrm{j}2\pi f\frac{Rd,1}{c}}\boldsymbol{\alpha}(\theta_d)\end{aligned} \tag{5.23}$$

接下来推导参考阵元接收到的反射波信号，即

$$\rho_1 s(t)\mathrm{e}^{-\mathrm{j}2\pi f\frac{Rs,1}{c}},R_{s,1}=\sqrt{(h_t+h_1)^2+R^2}\approx\bar{R}+h_1\sin\theta_s=R_{d,1}+\Delta R_1 \tag{5.24}$$

其中，θ_s 为目标反射波的角度，ΔR_1 为波程差，其值等于 $\Delta R_1=2\dfrac{h_1h_t}{R}$。

从上往下数第二个阵元的反射波信号为

$$\rho_2 s(t)\mathrm{e}^{-\mathrm{j}2\pi f\frac{Rs,2}{c}},R_{s,2}=\sqrt{(h_t+h_2)^2+R^2}\approx\bar{R}+h_2\sin\theta_s \tag{5.25}$$

第 N 个阵元的反射波信号为

$$\rho_N s(t)\mathrm{e}^{-\mathrm{j}2\pi f\frac{Rs,N}{c}},R_{s,N}=\sqrt{(h_t+h_N)^2+R^2}\approx\bar{R}+h_N\sin\theta_s \tag{5.26}$$

所以反射波信号矢量可以写成：

$$\begin{aligned}\boldsymbol{x}_s(t)&=[\rho_1 s(t)\mathrm{e}^{-\mathrm{j}2\pi f\frac{Rs,1}{c}}\ \rho_2 s(t)\mathrm{e}^{-\mathrm{j}2\pi f\frac{Rs,2}{c}}\cdots \rho_N s(t)\mathrm{e}^{-\mathrm{j}2\pi f\frac{Rs,N}{c}}]^{\mathrm{T}}\\&=s(t)\mathrm{e}^{-\mathrm{j}2\pi f\frac{\bar{R}}{c}}[\rho_1\mathrm{e}^{\mathrm{j}2\pi f\frac{-h1\sin\theta_s}{c}}\ \rho_2\mathrm{e}^{\mathrm{j}2\pi f\frac{-h2\sin\theta_s}{c}}\cdots \rho_N\mathrm{e}^{-\mathrm{j}2\pi f\frac{-hN\sin\theta_s}{c}}]^{\mathrm{T}}\\&=s(t)\mathrm{e}^{-\mathrm{j}2\pi f\frac{\bar{R}+(ha+(N-1)d)\sin\theta_s}{c}}[\rho_1\ \rho_2\mathrm{e}^{-\mathrm{j}2\pi f\frac{d\sin\theta_s}{c}}\cdots \rho_N\mathrm{e}^{-\mathrm{j}2\pi f\frac{(N-1)d\sin\theta_s}{c}}]^{\mathrm{T}}\\&=s(t)\mathrm{e}^{-\mathrm{j}2\pi f\frac{Rs,1}{c}}\boldsymbol{\rho}\odot[1\ \mathrm{e}^{-\mathrm{j}2\pi f\frac{d\sin\theta_s}{c}}\cdots \mathrm{e}^{-\mathrm{j}2\pi f\frac{(N-1)d\sin\theta_s}{c}}]^{\mathrm{T}}\end{aligned}$$

$$=s(t)\mathrm{e}^{-\mathrm{j}2\pi f\frac{R_{s,1}}{c}}\boldsymbol{\rho}\odot\boldsymbol{\alpha}(\theta_s) \tag{5.27}$$

则整个阵列的接收数据为

$$\begin{aligned}\boldsymbol{x}(t)&=\boldsymbol{x}_d(t)+\boldsymbol{x}_s(t)+\boldsymbol{n}(t)\\&=s(t)\mathrm{e}^{-\mathrm{j}2\pi f\frac{R_{d,1}}{c}}\boldsymbol{\alpha}(\theta_d)+s(t)\mathrm{e}^{-\mathrm{j}2\pi f\frac{R_{s,1}}{c}}\boldsymbol{\rho}\odot\boldsymbol{\alpha}(\theta_s)+\boldsymbol{n}(t)\\&=[\boldsymbol{\alpha}(\theta_d)+\mathrm{e}^{-\mathrm{j}2\pi f\frac{R_{s,1}-R_{d,1}}{c}}\boldsymbol{\rho}\odot\boldsymbol{\alpha}(\theta_s)]s(t)\mathrm{e}^{-\mathrm{j}2\pi f\frac{R_{d,1}}{c}}+\boldsymbol{n}(t)\\&=[\boldsymbol{\alpha}(\theta_d)+\mathrm{e}^{-\mathrm{j}2\pi f\frac{2h_1h_t}{cR}}\boldsymbol{\rho}\odot\boldsymbol{\alpha}(\theta_s)]s(t)\mathrm{e}^{-\mathrm{j}2\pi f\frac{R_{d,1}}{c}}+\boldsymbol{n}(t)\end{aligned} \tag{5.28}$$

注意到上述平坦阵地的信号模型与传统的信号模型不一致的地方在于反射系数，上述为一个矢量，而传统反射系数是一个标量，即反射点的位置被笼统地看成一个反射点，如图 5.10 所示。那么反射系数即为$[\rho_1=\rho_2=\cdots=\rho_N=\rho]$。实际上这里进行了简化，可以看到各反射点的间距为 $d/\tan\theta_s$，假设 $d=1$ m，反射角度 $\theta_s=2°$，则间距为 28.6 m，可以看到各阵元之间的反射点间距还是比较大的，这样当地面介质不太均匀的时候，传统算法的反射系数设置为一个值并不精确，会影响测高效果。

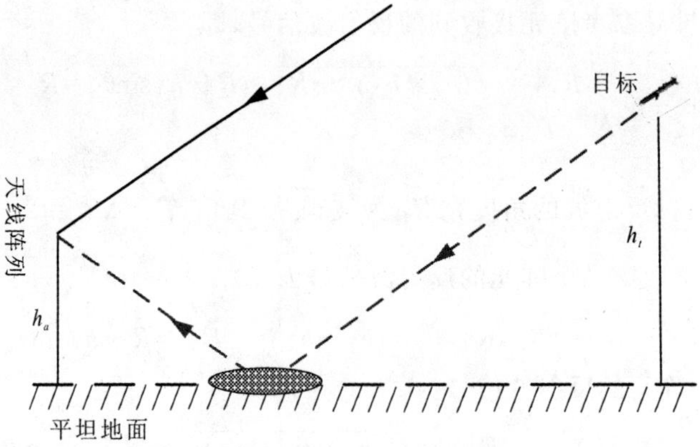

图 5.10　平坦阵地反射信号简化模型

2.传统的起伏阵地测高模型

首先介绍传统起伏阵地的信号模型，认为只有一个起伏反射点，实际上相当于算法做了两个假设：一是与平坦阵地中一致认为反射点只集中在一个物理几何点；二是认为起伏反射点只有一个值，这相当于起伏阵地的简化版信号模型，如图 5.11 所示。

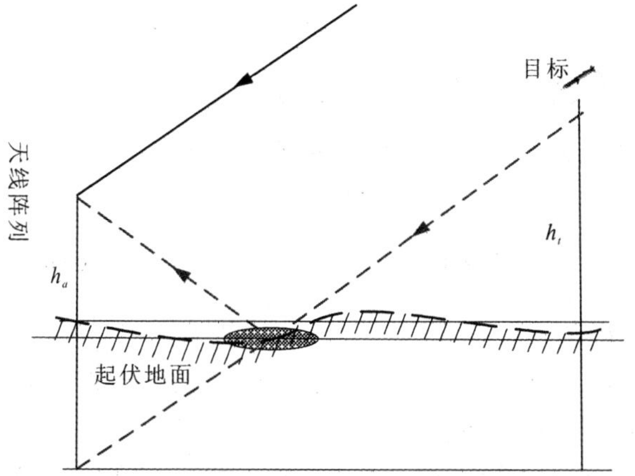

图 5.11　起伏阵地反射信号简化模型

对于起伏阵地模型，其直达波的接收信号与上述平坦阵地是一致的，则有

$$x_d(t)=s(t)\mathrm{e}^{-\mathrm{j}2\pi f\frac{Rd,1}{c}}a(\theta_d) \tag{5.29}$$

对于反射信号则需要根据起伏阵地来确定，将对应于多个阵元的反射点归结为同一个反射点。所以首先需要定义反射点高度 h_s，定义反射点在阵地水平面之上为负数，若反射点在阵地水平面之下为正数，则可看到反射点高度为 h_a+h_s，其中 h_a 为天线的高度(实际上在此简化模型当中，以前的文献和算法并不明确到底是哪个阵元的高度)。那么反射点的水平距离为$(h_a+h_s)\tan(\theta_s)$，反射系数的计算就需要用到该点的介电常数和电导率。

下面推导反射波的信号模型。第 i 个阵元的反射波信号为

$$\rho_i s(t)\mathrm{e}^{-\mathrm{j}2\pi f\frac{\bar{R}_{s,i}}{c}},\bar{R}_{s,i}=\sqrt{(h_t+h_i+2h_{s,i})^2+R^2}\approx\bar{R}+(h_i+2h_{s,i})\sin\theta_s \tag{5.30}$$

这里需要重点强调的一点是，该模型将反射系数设置为一个值，即$[\rho_1=\rho_2=\cdots=\rho_N=\rho]$，反射点起伏归结为一个值，即$[h_{s,1}=h_{s,2}=\cdots=h_{s,N}=h_s]$。整个阵列的反射信号矢量为

$$\begin{aligned}\boldsymbol{x}_{\mathrm{s,coarse}}(t)&=[\rho s(t)\mathrm{e}^{-\mathrm{j}2\pi f\frac{\bar{R}_{s,1}}{c}}\ \rho s(t)\mathrm{e}^{-\mathrm{j}2\pi f\frac{\bar{R}_{s,2}}{c}}\cdots\ \rho s(t)\mathrm{e}^{-\mathrm{j}2\pi f\frac{\bar{R}_{s,N}}{c}}]^{\mathrm{T}}\\&=\rho s(t)\mathrm{e}^{-\mathrm{j}2\pi f\frac{c}{}}[\mathrm{e}^{\mathrm{j}2\pi f\frac{-(h1+2hs)\sin\theta_s}{c}}\ \mathrm{e}^{\mathrm{j}2\pi f\frac{-(h2+2hs)\sin\theta_s}{c}}\cdots\ \mathrm{e}^{-\mathrm{j}2\pi f\frac{-(hN+2hs)\sin\theta_s}{c}}]^{\mathrm{T}}\\&=\rho s(t)\mathrm{e}^{-\mathrm{j}2\pi f\frac{\bar{R}+(h1+2hs)\sin\theta}{c}}[1\ \mathrm{e}^{-\mathrm{j}2\pi f\frac{d\sin\theta_s}{c}}\cdots\ \mathrm{e}^{-\mathrm{j}2\pi f\frac{(N-1)d\sin\theta_s}{c}}]^{\mathrm{T}}\end{aligned}$$

$$=\rho s(t)\mathrm{e}^{-\mathrm{j}2\pi f\frac{\bar{R}+(h_1+2h_s)\sin\theta}{c}}\left[1\ \mathrm{e}^{-\mathrm{j}2\pi f\frac{d\sin\theta_s}{c}}\cdots,\mathrm{e}^{-\mathrm{j}2\pi f\frac{(N-1)d\sin\theta_s}{c}}\right]^{\mathrm{T}}$$

$$=\rho s(t)\mathrm{e}^{-\mathrm{j}2\pi f\frac{\bar{R}+(h_1+2h_s)\sin\theta_s}{c}}\boldsymbol{\alpha}(\theta_s) \tag{5.31}$$

所以整个阵列的直达波加反射波的阵列接收数据为

$$\boldsymbol{x}(t)=\boldsymbol{x}_d(t)+\boldsymbol{x}_{\mathrm{s,coarse}}(t)+\boldsymbol{n}(t)$$

$$=s(t)\mathrm{e}^{-\mathrm{j}2\pi f\frac{Rd,1}{c}}\boldsymbol{\alpha}(\theta_d)+s(t)\mathrm{e}^{-\mathrm{j}2\pi f\frac{\bar{R}+(h_1+2h_s)\sin\theta_s}{c}}\rho\boldsymbol{\alpha}(\theta_s)+\boldsymbol{n}(t)$$

$$=\left[\boldsymbol{\alpha}(\theta_d)+\mathrm{e}^{-\mathrm{j}2\pi f\frac{\bar{R}+(h_1+2h_s)\sin\theta_s-Rd,1}{c}}\rho\boldsymbol{\alpha}(\theta_s)\right]s(t)\mathrm{e}^{-\mathrm{j}2\pi f\frac{Rd,1}{c}}+\boldsymbol{n}(t)$$

$$=\left[\boldsymbol{\alpha}(\theta_d)+\mathrm{e}^{-\mathrm{j}2\pi f\frac{2(h_1+2h_s)h_t}{cR}}\rho\boldsymbol{a}(\theta_s)\right]s(t)\mathrm{e}^{-\mathrm{j}2\pi f\frac{Rd,1}{c}}+\boldsymbol{n}(t) \tag{5.32}$$

3.精细化起伏阵地测高模型

事实上，h_t 与阵列天线并不在一个水平线上，反射点位置不一致，$h_{s,i}$ 也不在同一水平线上，其值是不相等的。起伏阵地反射信号模型的精细化结构如图 5.12 所示。下面来分析精细化信号模型，并根据反射点高度计算出反射点位置。

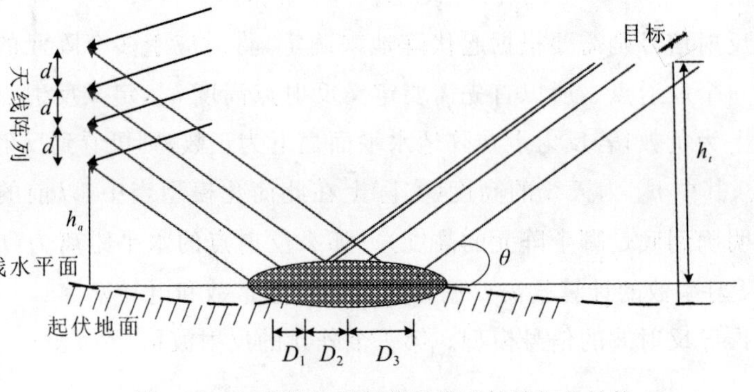

图 5.12　起伏阵地反射信号模型的精细化结构

由图 5.12 可以看到利用现有算法直达波的波程也与传统算法有所不同，因为要用到 h_t 的信息，所以需要注意的是这里的 h_t 为目标离地面的高度。可以看到反射点的分布并不均匀，即 D_1，D_2，D_3 并不相等。下面详细推导其信号模型。参考阵元接收到的直达波信号为

$$s(t)\mathrm{e}^{-\mathrm{j}2\pi f\frac{\bar{\bar{R}}_{d,1}}{c}},\bar{\bar{R}}_{d,1}=\sqrt{(h_t-h_1)^2+R^2}\approx\bar{R}-h_1\sin\theta_d \tag{5.33}$$

第二个阵元的直达波信号为

$$s(t)\mathrm{e}^{-\mathrm{j}2\pi f\frac{\bar{\bar{R}}_{d,2}}{c}},\bar{\bar{R}}_{d,2}=\sqrt{(h_t-h_2)^2+R^2}\approx\bar{R}-h_2\sin\theta_d \tag{5.34}$$

第 N 个阵元的直达波信号为

$$s(t)\mathrm{e}^{-\mathrm{j}2\pi f\frac{\bar{\bar{R}}_{d,N}}{c}},\bar{\bar{R}}_{d,N}=\sqrt{(h_t-h_N)^2+R^2}\approx\bar{R}-h_N\sin\theta_d \tag{5.35}$$

所以直达波信号矢量可以写成

$$\begin{aligned}
\bar{\bar{\boldsymbol{x}}}_d(t)&=[s(t)\mathrm{e}^{-\mathrm{j}2\pi f\frac{\bar{\bar{R}}_{d,1}}{c}}\ s(t)\mathrm{e}^{-\mathrm{j}2\pi f\frac{\bar{\bar{R}}_{d,2}}{c}}\cdots s(t)\mathrm{e}^{-\mathrm{j}2\pi f\frac{\bar{\bar{R}}_{d,N}}{c}}]^{\mathrm{T}}\\
&=s(t)\mathrm{e}^{-\mathrm{j}2\pi f\frac{\bar{R}\sin\theta d}{c}}[\mathrm{e}^{\mathrm{j}2\pi f\frac{h1\sin\theta d}{c}}\ \mathrm{e}^{\mathrm{j}2\pi f\frac{h2\sin\theta d}{c}}\cdots\mathrm{e}^{-\mathrm{j}2\pi f\frac{hN\sin\theta d}{c}}]^{\mathrm{T}}\\
&=s(t)\mathrm{e}^{-\mathrm{j}2\pi f\frac{\bar{R}-h1\sin\theta d}{c}}[1\ \mathrm{e}^{-\mathrm{j}2\pi f\frac{d\sin\theta d}{c}}\cdots\mathrm{e}^{-\mathrm{j}2\pi f\frac{(N-1)d\sin\theta d}{c}}]^{\mathrm{T}}\\
&=s(t)\mathrm{e}^{-\mathrm{j}2\pi f\frac{\bar{\bar{R}}=d,1}{c}}[1\ \mathrm{e}^{-\mathrm{j}2\pi f\frac{d\sin\theta d}{c}}\cdots\mathrm{e}^{-\mathrm{j}2\pi f\frac{(N-1)d\sin\theta d}{c}}]^{\mathrm{T}}\\
&=s(t)\mathrm{e}^{-\mathrm{j}2\pi f\frac{\bar{\bar{R}}_{d,1}}{c}}\boldsymbol{\alpha}(\theta_d)
\end{aligned} \tag{5.36}$$

参考阵元接收到的反射波信号为

$$\begin{aligned}
\rho_1 s(t)\mathrm{e}^{-\mathrm{j}2\pi f\frac{\bar{\bar{R}}_{s,1}}{c}},\bar{\bar{R}}_{s,1}&=\sqrt{[h_t+(h_1+2h_{s,1})]^2+R^2}\\
&\approx\bar{R}+(h_1+2h_{s,1})\sin\theta_s\\
&=\bar{\bar{R}}_{d,1}+\Delta R_1
\end{aligned} \tag{5.37}$$

其中，波程差 $\Delta R_1=2\dfrac{(h_1+2h_{s,1})h_t}{R}$，$h_t-h_1=R\tan\theta_d$，$h_t+h_1=R\tan\theta_s$。

第二个阵元的反射波信号为

$$\begin{aligned}
\rho_2 s(t)\mathrm{e}^{-\mathrm{j}2\pi f\frac{\bar{\bar{R}}_{s,2}}{c}},\bar{\bar{R}}_{s,2}&=\sqrt{[h_t+(h_2+2h_{s,2})]^2+R^2}\\
&\approx\bar{R}+(h_2+2h_{s,2})\sin\theta_s\\
&=R_{d,2}+2\frac{(h_2+2h_{s,2})h_t}{R}
\end{aligned} \tag{5.38}$$

第 N 个阵元的反射波信号为

$$\begin{aligned}
\rho_N s(t)\mathrm{e}^{-\mathrm{j}2\pi f\frac{\bar{\bar{R}}_{s,N}}{c}},\bar{\bar{R}}_{s,N}&=\sqrt{[h_t+(h_N+2h_{s,N})]^2+R^2}\\
&\approx\bar{R}+(h_N+2h_{s,N})\sin\theta_s\\
&=R_{d,N}+2\frac{(h_N+2h_{s,N})h_t}{R}
\end{aligned} \tag{5.39}$$

所以反射波信号矢量可以写成

$$
\begin{aligned}
\bar{\bar{x}}_s(t)&=[\rho_1 s(t)\mathrm{e}^{-\mathrm{j}2\pi f\frac{\bar{\bar{R}}_{s,1}}{c}}\ \rho_2 s(t)\mathrm{e}^{-\mathrm{j}2\pi f\frac{\bar{\bar{R}}_{s,2}}{c}}\cdots \rho_N s(t)\mathrm{e}^{-\mathrm{j}2\pi f\frac{\bar{\bar{R}}_{s,N}}{c}}]^{\mathrm{T}}\\
&=s(t)\mathrm{e}^{-\mathrm{j}2\pi f\frac{}{c}}[\rho_1\mathrm{e}^{\mathrm{j}2\pi f\frac{-(h_1+2h_{s,1})\sin\theta_s}{c}}\ \rho_2\mathrm{e}^{\mathrm{j}2\pi f\frac{-(h_2+2h_{s,2})\sin\theta_s}{c}}\cdots \rho_N\mathrm{e}^{-\mathrm{j}2\pi f\frac{-(h_N+2h_{s,N})\sin\theta_s}{c}}]^{\mathrm{T}}\\
&=s(t)\mathrm{e}^{-\mathrm{j}2\pi f\frac{\bar{R}+h_1\sin\theta_s}{c}}\\
&\quad[\rho_1\mathrm{e}^{\mathrm{j}2\pi f\frac{-2h_{s,1}\sin\theta_s}{c}}\ \rho_2\mathrm{e}^{\mathrm{j}2\pi f\frac{-2h_{s,2}\sin\theta_s}{c}}\mathrm{e}^{-\mathrm{j}2\pi f\frac{d\sin\theta_s}{c}}\cdots\rho_N\mathrm{e}^{\mathrm{j}2\pi f\frac{-2h_{s,N}\sin\theta_s}{c}}\mathrm{e}^{-\mathrm{j}2\pi f\frac{(N-1)d\sin\theta_s}{c}}]^{\mathrm{T}}\\
&=s(t)\mathrm{e}^{-\mathrm{j}2\pi f\frac{\bar{R}+h_1\sin\theta_s}{c}}\\
&\quad[\rho_1\mathrm{e}^{\mathrm{j}2\pi f\frac{-2h_{s,1}\sin\theta_s}{c}}\ \rho_2\mathrm{e}^{\mathrm{j}2\pi f\frac{-2h_{s,2}\sin\theta_s}{c}}\mathrm{e}^{-\mathrm{j}2\pi f\frac{d\sin\theta_s}{c}}\cdots\rho_N\mathrm{e}^{\mathrm{j}2\pi f\frac{-2h_{s,N}\sin\theta_s}{c}}\mathrm{e}^{-\mathrm{j}2\pi f\frac{(N-1)d\sin\theta_s}{c}}]^{\mathrm{T}}
\end{aligned}
\tag{5.40}
$$

其中，

$$
\boldsymbol{\rho}\Delta=[\rho_1\ \rho_2\cdots\rho_N]^{\mathrm{T}} \tag{5.41}
$$

$$
\boldsymbol{\alpha}(h_{s,1},h_{s,2}\cdots h_{s,N})=[\mathrm{e}^{\mathrm{j}2\pi f\frac{-2h_{s,1}\sin\theta_s}{c}}\ \mathrm{e}^{\mathrm{j}2\pi f\frac{-2h_{s,2}\sin\theta_s}{c}}\cdots\mathrm{e}^{\mathrm{j}2\pi f\frac{-2h_{s,N}\sin\theta_s}{c}}]^{\mathrm{T}} \tag{5.42}
$$

$$
\bar{\bar{\boldsymbol{x}}}_s(t)=s(t)\mathrm{e}^{-\mathrm{j}2\pi f\frac{\bar{R}+h_1\sin\theta_s}{c}}\boldsymbol{\rho}\odot\boldsymbol{\alpha}(h_{s,1},h_{s,2}\cdots h_{s,N})\odot\boldsymbol{\alpha}(\theta_s) \tag{5.43}
$$

所以整个阵列的信号模型为

$$
\begin{aligned}
\boldsymbol{x}(t)&=\bar{\bar{\boldsymbol{x}}}_d(t)+\bar{\bar{\boldsymbol{x}}}_s(t)+\boldsymbol{n}(t)\\
&=s(t)\mathrm{e}^{-\mathrm{j}2\pi f\frac{\bar{\bar{R}}_{d,1}}{c}}\boldsymbol{\alpha}(\theta_d)+s(t)\mathrm{e}^{-\mathrm{j}2\pi f\frac{\bar{R}+h_1\sin\theta_s}{c}}\boldsymbol{\rho}\odot\boldsymbol{a}(h_{s,1},h_{s,2}\cdots h_{s,N})\odot\boldsymbol{a}(\theta_s)+\boldsymbol{n}(t)\\
&=[\boldsymbol{a}(\theta_d)+\mathrm{e}^{-\mathrm{j}2\pi f\frac{(\bar{R}+h_1\sin\theta_s)-\bar{\bar{R}}_{d,1}}{c}}\boldsymbol{\rho}\odot\boldsymbol{a}(h_{s,1},h_{s,2}\cdots h_{s,N})\odot\boldsymbol{a}(\theta_s)]s(t)\mathrm{e}^{-\mathrm{j}2\pi f\frac{\bar{\bar{R}}_{d,1}}{c}}+\boldsymbol{n}(t)\\
&=[\boldsymbol{\alpha}(\theta_d)+\mathrm{e}^{-\mathrm{j}2\pi f\frac{2h_1h_t}{cR}}\boldsymbol{\rho}\odot\boldsymbol{a}(h_{s,1},h_{s,2}\cdots h_{s,N})\odot\boldsymbol{a}(\theta_s)]s(t)\mathrm{e}^{-\mathrm{j}2\pi f\frac{R_{d,1}}{c}}+\boldsymbol{n}(t)
\end{aligned}
\tag{5.44}
$$

需要讨论的是反射点的水平距离，因为反射点的位置直接决定了反射系数。从图 5.12 中可以看到每个阵元的反射点是不一致的，反射点是入射的反射波仰角 θ_s 和反射点高度 $h_{s,i}$ 的函数，可通过简单的三角函数计算得到每个阵元的反射点水平距离为：$(h_i+h_{s,i})\tan\theta_s$。式(5.44)中的波程差被定义为：

$$
\delta=2\pi f\ \frac{2h_1h_t}{cR} \tag{5.45}
$$

由此可见，由起伏地形定义的波程差与非起伏传统地形波程差是一致的，所有地形起伏均被计算到起伏导向矢量 $\boldsymbol{\alpha}(h_{s,1},h_{s,2},\cdots,h_{s,N})$ 中。

4.复杂地形 MIMO 雷达测高精确模型

上面已经建立了复杂阵地的信号接收模型，下面将其推广到米波 MIMO

阵列雷达测高中。米波 MIMO 雷达的一个目标对应四条传输路径的信号模型，如图 5.13 所示。

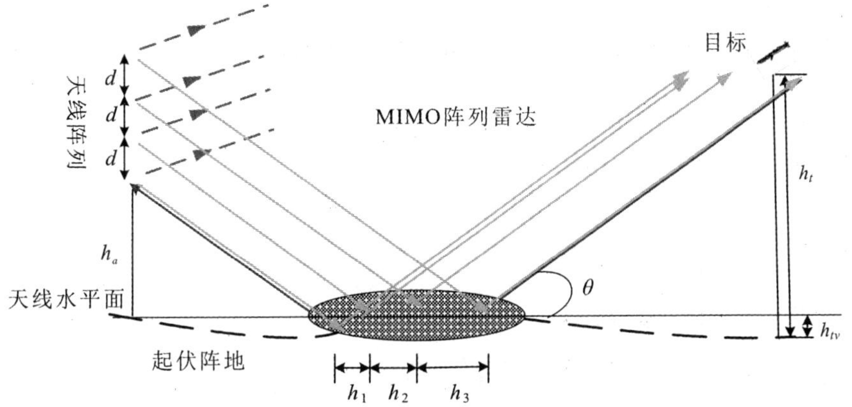

图 5.13　米波 MIMO 雷达低仰角测高复杂阵地反射模型示意图

则可将式(5.44)中的两条路径扩展为四条路径，对于 MIMO 雷达其匹配滤波之后的信号模型为两个信号的 kroneker 积，这在以往的米波 MIMO 雷达中已经进行过推导[158]，下面将其扩展成复杂地形的形式：

$$\begin{aligned}\boldsymbol{x}_{\text{mimo}}(t)=&[\boldsymbol{\alpha}_{\text{t}}(\theta_d)+\mathrm{e}^{-\mathrm{j}\delta}\boldsymbol{\rho}(h_{s,i})\odot\boldsymbol{\alpha}_{terrain}(h_{s,i})\odot\boldsymbol{\alpha}_{\text{t}}(\theta_s)]\otimes\\&[\boldsymbol{\alpha}_r(\theta_d)+\mathrm{e}^{-\mathrm{j}\delta}\boldsymbol{\rho}(h_{s,i})\odot\boldsymbol{\alpha}_{\text{terrain}}(h_{s,i})\odot\boldsymbol{\alpha}_r(\theta_s)]s(t)+\boldsymbol{n}_{\text{mimo}}(t)\end{aligned}\tag{5.46}$$

在这个信号模型中，假设发射和接收阵元共用，即收发共置，则

$$\left.\begin{aligned}\boldsymbol{\alpha}_{\text{t}}(\theta_d)=\boldsymbol{\alpha}_r(\theta_d)=\boldsymbol{\alpha}(\theta_d)\\\boldsymbol{\alpha}_{\text{t}}(\theta_s)=\boldsymbol{\alpha}_r(\theta_s)=\boldsymbol{\alpha}(\theta_s)\end{aligned}\right\}\tag{5.47}$$

且通常假设噪声 $\boldsymbol{n}_{\text{mimo}}(t)\in\mathbb{C}^{M^2\times 1}$ 为白噪声。米波 MIMO 雷达的测高任务就是根据式(5.46)的接收数据来计算其中的 θ_d 参数。

§5.2.3　最大似然和广义 MUSIC 算法的精确信号模型测高应用

通过上节分析可知，无需解相干类的超分辨算法，如最大似然算法和广义 MUSIC 算法在 MIMO 雷达测高得到充分使用，是两种行之有效的测高算法。下面推导这两种算法在精确地形模型下的米波 MIMO 雷达中的应用。

1.最大似然

首先将式(5.46)中 MIMO 雷达接收数据中的信号源和导向矢量重新组合成如下形式：

$$[\boldsymbol{\alpha}_t(\theta_d)+e^{-j\delta}\boldsymbol{\rho}(h_{s,i})\odot\boldsymbol{\alpha}_t(h_{s,i})\odot\boldsymbol{\alpha}_t(\theta_s)]\otimes[\boldsymbol{\alpha}_r(\theta_d)+e^{-j\delta}\boldsymbol{\rho}(h_{s,i})\odot\boldsymbol{\alpha}_r(h_{s,i})\odot\boldsymbol{\alpha}_r(\theta_s)]s(t)=$$

$$[\boldsymbol{\alpha}_t(\theta_d)\boldsymbol{\rho}(h_{s,i})\odot\boldsymbol{\alpha}_t(h_{s,i})\odot\boldsymbol{\alpha}_t(\theta_s)]\otimes[\boldsymbol{\alpha}_r(\theta_d)\boldsymbol{\rho}(h_{s,i})\odot\boldsymbol{\alpha}_r(h_{s,i})\odot\boldsymbol{\alpha}_r(\theta_s)]\begin{bmatrix}1\\e^{j\delta}\\e^{j\delta}\\e^{j2\delta}\end{bmatrix}s(t) \tag{5.48}$$

则可将式(5.46)中的接收数据写成如下形式：

$$\boldsymbol{x}_{\text{mimo}}(t)=\boldsymbol{A}_{\text{mimo}}(\theta_d,\theta_s,h_{s,i})\bar{\boldsymbol{s}}(t)+\boldsymbol{n}_{\text{mimo}}(t) \tag{5.49}$$

其中

$$\boldsymbol{A}(\theta_d,\theta_s,h_{s,i})=[\boldsymbol{\alpha}_t(\theta_d)\boldsymbol{\rho}(h_{s,i})\odot\boldsymbol{\alpha}_t(h_{s,i})\odot\boldsymbol{\alpha}_t(\theta_s)]\otimes [\boldsymbol{\alpha}_r(\theta_d)\boldsymbol{\rho}(h_{s,i})\odot\boldsymbol{\alpha}_r(h_{s,i})\odot\boldsymbol{\alpha}_r(\theta_s)]\in\mathbb{C}^{M^2\times 4} \tag{5.50}$$

等效信号矢量为

$$\bar{\boldsymbol{s}}(t)=\begin{bmatrix}1\\e^{j\delta}\\e^{j\delta}\\e^{j2\delta}\end{bmatrix}s(t) \tag{5.51}$$

则根据上节分析可知，最大似然估计准则可表示为

$$[\theta_d\ \theta_s,h_{s,i}]=-\underset{[\theta_d\theta_s,h_{s,i}]}{\operatorname{argmax}}\operatorname{Tr}[\boldsymbol{P}_{\boldsymbol{A}_{\text{mimo}}(\theta_d,\theta_s,h_{s,i})}\hat{\boldsymbol{R}}_{\text{mimo}}] \tag{5.52}$$

其中，$\boldsymbol{A}_{\text{mimo}}(\theta_d,\theta_s,h_{s,i})$为导向矢量矩阵，$\hat{\boldsymbol{R}}_{\text{mimo}}$为阵列输出信号的协方差矩阵估计值，$\boldsymbol{P}_{\boldsymbol{A}_{\text{mimo}}(\theta_d,\theta_s,h_{s,i})}$为投影到导向矢量矩阵 $\boldsymbol{A}_{\text{mimo}}(\theta_d,\theta_s,h_{s,i})$的列向量张成空间的投影矩阵，其值等于：

$$\boldsymbol{P}_{\boldsymbol{A}_{\text{mimo}}(\theta_d,\theta_s,h_{s,i})}=\boldsymbol{A}_{\text{mimo}}(\theta_d,\theta_s,h_{s,i})[\boldsymbol{A}_{\text{mimo}}(\theta_d,\theta_s,h_{s,i})^{\text{H}}\boldsymbol{A}_{\text{mimo}}(\theta_d,\theta_s,h_{s,i})]^{-1} \boldsymbol{A}_{\text{mimo}}(\theta_d,\theta_s,h_{s,i})^{\text{H}} \tag{5.53}$$

通过上述定义，可以看出对应于本文的 MIMO 阵列雷达，式(5.52)中最大似然 $[\theta_d\ \theta_s,h_{s,i}]$为信源的直达波、反射波和阵地起伏参数，式(5.52)中的信号协方差矩阵为

$$\hat{\boldsymbol{R}}_{\text{mimo}}=E\{\boldsymbol{x}_{\text{mimo}}(t)\boldsymbol{x}_{\text{mimo}}(t)^{\text{H}}\} \tag{5.54}$$

2.广义 MUSIC

根据上面的分析得知传统 MUSIC 算法无法直接应用，而广义 MUSIC 可直接应用。对 MIMO 阵列雷达匹配滤波后接收数据协方差矩阵$\hat{\boldsymbol{R}}_{\text{mimo}}$进行特征分解，得到噪声子空间 $\boldsymbol{E}_n$ 和信号子空间 $\boldsymbol{E}_s$，根据上节推导公式，广义 MUSIC 谱为

$$P(\boldsymbol{\theta})=\frac{\det[\boldsymbol{A}_{\text{mimo}}(\theta_d,\theta_s,h_{s,i})^{\mathrm{H}}\boldsymbol{A}_{\text{mimo}}(\theta_d,\theta_s,h_{s,i})]}{\det[\boldsymbol{A}_{\text{mimo}}(\theta_d,\theta_s,h_{s,i})^{\mathrm{H}}\boldsymbol{E}_n\boldsymbol{E}_n^{\mathrm{H}}\boldsymbol{A}_{\text{mimo}}(\theta_d,\theta_s,h_{s,i})]} \tag{5.55}$$

3.降维处理

最大似然的式(5.52)和广义 MUSIC 的式(5.55)均涉及多维搜索，首先可利用直达波和反射波的几何关系来降维。另外，在下面的仿真中，做一些假设来验证本文算法的有效性，一是假设在需要计算的仰角上，其地形起伏高度已知，即参数 $h_{s,i}$，$i=1,\cdots,M$已知，二是在计算反射系数时，反射系数中的介电常数 ε_r 和表面物质传导率σ_e 已知，反射系数的计算公式在上节的式(5.3)中定义。

4.仿真结果分析

米波 MIMO 阵列雷达发射阵元数 $M=10$，阵元间距为半波长。入射频率为 300 MHz，入射波长 $\lambda=1$ m，目标直达波角度设为 $\theta_d=6°$，反射角可按式(5.13)计算。天线高度 $h_a=5$ m，目标高度 $h_t=3\ 000$ m。

复杂地形设置如下：反射系数设置为两部分，前半阵列被设置为介电常数和表面材料电导率，$\varepsilon_r=80$，$\sigma_e=0.2$；后半阵列被设置为介电常数和表面材料电导率，$\varepsilon_r=75$，$\sigma_e=0.5$。反射区的复杂地形起伏高度设置为 $\boldsymbol{h}_{s,i,i=1,\cdots,M}=[0.1\ 0.2\ 0.3\ 0.4\ 0.5\ 0.5\ 0.4\ 0.3\ 0.2\ 0.1]^{\mathrm{T}}$。

仿真一：空间谱。

信噪比设为 SNR=20 dB。快照数设置为 10。图 5.14 和图 5.15 分别给出了基于最大似然估计和广义 MUSIC 的地形匹配算法，并分别给出了复杂地形引起的地形失配的仿真结果。图中，基于最大似然和广义 MUSIC 的地形匹配算法分别用“地形匹配最大似然”和“地形匹配广义 MUSIC”标记。基于最大似然和广义 MUSIC 的地形失配算法分别用“地形失配最大似然”和“地形失配广义 MUSIC”进行标记。可以看出，匹配算法能够正确估计出实际角度值，但失配算法的谱峰值发生了偏移。

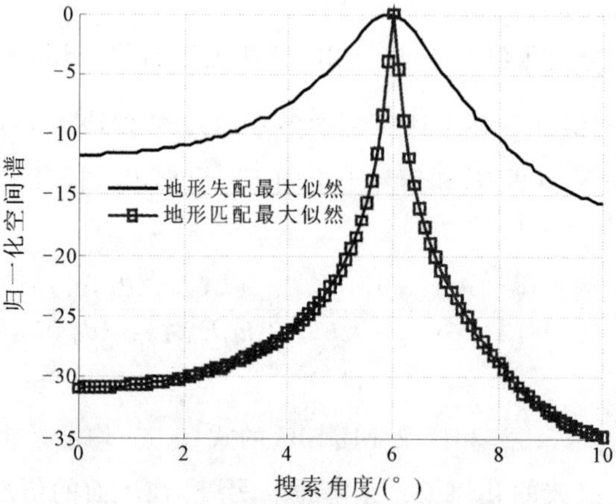

图 5.14　最大似然估计结果

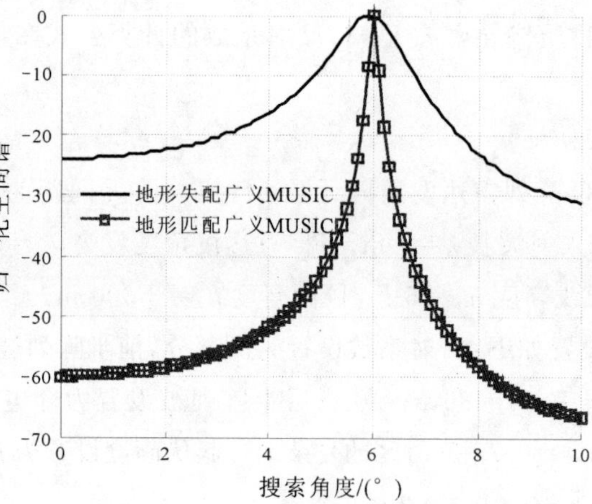

图 5.15　广义 MUSIC 结果

仿真二：高程和高度估计随信噪比变化的 RMSE 结果。

估计结果的均方根误差（RMSE）定义为

$$\text{RMSE}=\sqrt{\frac{1}{\text{Monte}}\sum_{p=1}^{\text{Monte}}E[(\hat{\alpha}_p-\alpha_p)^2]} \tag{5.56}$$

其中，$\hat{\alpha}=\hat{\theta}_d$ 为直达波角度估计值，$\alpha=\theta_d$ 为实际目标角度值。估计目标高度时，$\hat{\alpha}=\hat{h}_t$，$\alpha=h_t$，是目标高度的实际值。在 1 000 个蒙特卡洛实验中，快照数设置为 10。信噪比从 0 变化到 25 dB。图 5.16 和图 5.17 给出了最大似然法以及广义 MUSIC 算法的角度估计 RMSE 和高度测量 RMSE。从图中可以看出,这两种算法可以正确应用于 MIMO 雷达测高领域。匹配算法的估计精度随着信噪比的增加而降低,是一种无偏估计。失配算法的估计精度不会随着信噪比的增加而降低,是一种有偏估计。

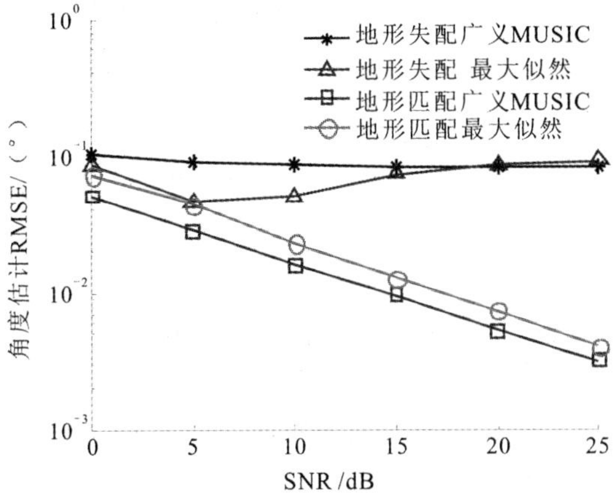

图 5.16　最大似然角估计 RMSE 与广义 MUSIC 算法

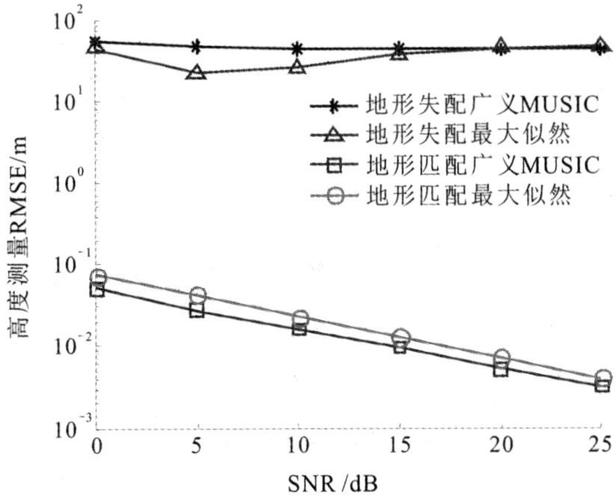

图 5.17　最大似然测高 RMSE 与广义 MUSIC 算法

仿真三:角度和高度估计的 RMSE 结果随快照数的变化。

信噪比设为 SNR=20 dB。快照的数量设置为 2 到 20。图 5.18 和图 5.19 给出了最大似然法和广义 MUSIC 算法下的角度估计 RMSE 和高度测量 RMSE 随快拍数的变化曲线。从图中可以看出,仿真三与仿真二的结论相似。

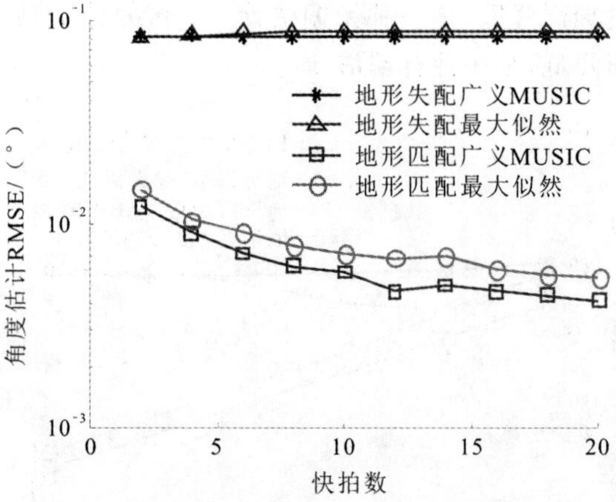

图 5.18 最大似然角估计 RMSE 与广义 MUSIC 算法

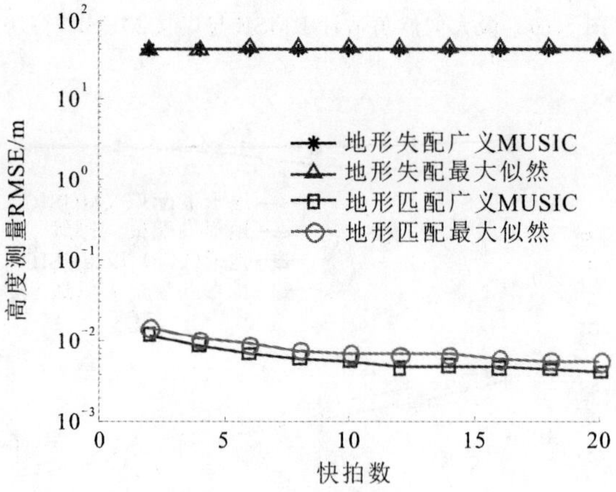

图 5.19 最大似然测高 RMSE 与广义 MUSIC 算法

§5.3 本章小结

本章研究了基于 BOMP 预处理的米波 MIMO 雷达的低仰角测高问题。首先对传统最大似然和广义 MUSIC 算法进行了回顾。接着提出了利用非解相干的稀疏恢复算法 BOMP 来粗测目标角度,然后缩小了传统算法的角度搜索范围，从而降低了计算量。事实上，若不考虑计算量，BOMP 亦可用。

针对米波雷达测高问题，笔者意识到阵地的反射信号模型是测高精度的关键点，指出平坦阵地模型反射点简化的问题，并给出反射点水平位置的计算，从而可根据表格得到各个阵元的反射系数，相比原有模型更加精确。另外，针对起伏阵地，传统反射模型失配严重，严重影响测高精度。本书精确给出反射的波程差，以及波程差与每个反射点高度的闭式计算，得到精确的反射导向矢量，并给出反射点位置的计算式。将其推广到 MIMO 雷达，给出了两种相应的测高算法。仿真结果表明，表面地形匹配算法优于地形失配估计算法。

第6章　基于波形与极化分集结合的米波极化 MIMO 雷达测高问题研究

§6.1　引　言

米波雷达测高问题从阵列的角度可分为常规阵列雷达、极化阵列雷达、MIMO 阵列雷达的测高问题等。下面从阵列的角度来分析其研究进展。

MIMO 雷达具有虚拟孔径扩展、抗干扰能力强、分辨率高、多目标跟踪能力强的优点，因此将 MIMO 雷达运用到米波雷达中解决米波雷达低仰角估计问题受到了国内外研究人员的广泛关注，这在上一章已经有了比较详细的分析。

另外，极化敏感阵列具有极化分集的优势，在空间受限情况下，极化通道能够比相同孔径的标量阵列提供更多的通道，通道间具有极化分集的优势，对两个邻近目标具有更好的分辨力。此外，对于相干目标来说，传统的相控阵标量阵列采用空间平滑来解相干，这将会导致实孔径变短。针对极化敏感阵列，可采用极化平滑的算法，利用极化分集对极化信息进行加权处理达到平滑的效果，来求解目标的相干性，不损失阵列孔径，提高相干目标的分辨性能。

本章将波形分集和极化分集相结合，研究米波极化 MIMO 雷达的测高问题，推导了米波极化 MIMO 雷达测高信号模型的闭式表达式，并提出基于 MUSIC 的两种测高算法。首先，详细推导如何将极化信息融入 MIMO 雷达中，给出极化参数对地面反射系数的影响，建立极化 MIMO 雷达测高信号模型。接着，进行一个重要推导，即对极化 MIMO 雷达信号模型进行适当的变形处理，为后续的常规 MUSIC 和广义 MUSIC 算法的应用做准备。然后详细推导广义 MUSIC 算法的应用。对于常规 MUSIC，提出用合成导向矢量的方法，并进行降维处理来降低算法复杂度。最后给出两种算法的计算量，并推导极化 MIMO 雷达的测高极限性能 CRB。

§6.2　极化 MIMO 雷达测高信号模型

假设一个收发共置的米波极化 MIMO 雷达系统，其发射阵元有 M 个，发射矢量信号为：$\boldsymbol{\varphi}(t)\in\mathbb{C}^{M\times1}$，该发射信号为正交信号(这是 MIMO 雷达区别于传统相控阵雷达的重要特征)，则有 $\int_0^{T_p}\boldsymbol{\varphi}(t)\boldsymbol{\varphi}(t)^{\mathrm{H}}\mathrm{d}t=\boldsymbol{I}_M$，其中，$\boldsymbol{I}_M$ 是大小为 M 的单位阵，T_p 为雷达一个脉冲持续时间，即脉冲宽度。对于连续波雷达则需要整个发射时间段的发射信号之间相互正交。阵列为半波长布阵的均匀线阵，在该系统的接收端，接收阵列采用六分量的电磁矢量传感器，故称该系统为米波极化 MIMO 雷达系统。假设极化 MIMO 雷达系统的低仰角反射区域是光滑平坦的反射面，则整个阵列如图 6.1 所示。

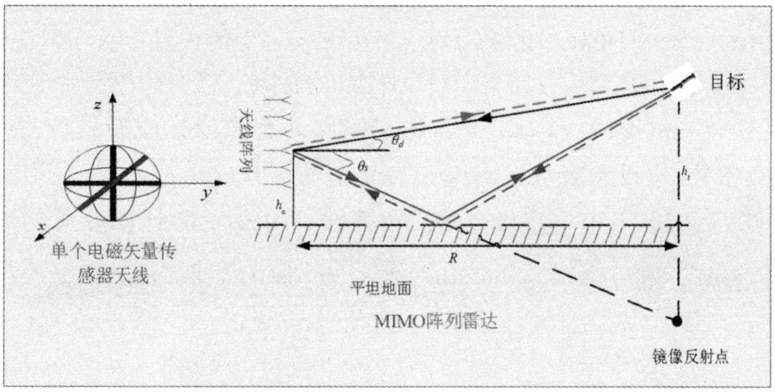

图 6.1　米波极化 MIMO 雷达低仰角测高镜面反射模型示意图

假设该系统的发射信号为水平极化信号，则其到达目标的信号可表示为

$$x(t)=[\boldsymbol{\alpha}_t(\theta_d)+\mathrm{e}^{\mathrm{j}\delta}\rho_h\boldsymbol{\alpha}_t(\theta_s)]^{\mathrm{T}}\boldsymbol{\varphi}(t) \tag{6.1}$$

其中，θ_d 为直达波角度，θ_s 为反射波角度，$\boldsymbol{\alpha}_t(\theta)$ 为发射阵列导向矢量，其值等于

$$\boldsymbol{\alpha}_t(\theta)=[1\ \exp(-\mathrm{j}2\pi d\sin(\theta)/\lambda)\ \cdots\ \exp(-\mathrm{j}2\pi(M-1)d\sin(\theta)/\lambda)]^{\mathrm{T}} \tag{6.2}$$

其中，θ 可以表示 θ_d 或者 θ_s，d 为阵元间隔，λ 为入射波长，$\delta=\dfrac{4\pi h_a h_t}{R\lambda}$ 为直达波和反射波的波程差所引起的相位差，其中 h_a 为天线高度(天线最高阵元位置)，h_t 为目标高度，R 为目标垂直投影到地面的点与雷达天线的水平距离。

通常来说，波程差小于米波雷达的一个距离分辨单元，故在距离上区分不了直达和反射波，这将造成信号消减、波瓣分裂，导致测高变得困难。ρ_h 为水平极化波的反射系数，见式(5.3)，其值如下：

$$\rho_h=\frac{\sin\theta_d-\sqrt{\varepsilon-\cos^2\theta_d}}{\sin\theta_d+\sqrt{\varepsilon-\cos^2\theta_d}} \tag{6.3}$$

其中，ε 为表面复介电常数，其数值可由相对介电常数 ε_r 和表面物质传导率 σ_e 表示：

$$\varepsilon=\varepsilon_r-\mathrm{j}60\lambda\sigma_e \tag{6.4}$$

在该系统的接收端，每个矢量阵元采用六分量的完全电磁形式，因为假设 MIMO 雷达系统收发共置，则接收阵列的空域导向矢量，其中，$\boldsymbol{\alpha}_r(\theta)$ 与发射导向矢量相等，为 $\boldsymbol{\alpha}_t(\theta)$，即 $\boldsymbol{\alpha}_r(\theta)=\boldsymbol{\alpha}_t(\theta)$。则整个阵列的接收数据可写成：

$$\begin{aligned}\boldsymbol{z}(t,\tau)&=\{\boldsymbol{\alpha}_r(\theta_d)\otimes[\boldsymbol{\Phi}(\theta_d)\boldsymbol{g}(\gamma,\eta)]+\mathrm{e}^{\mathrm{j}\delta}\boldsymbol{\alpha}_r(\theta_s)\otimes[\boldsymbol{\Phi}(\theta_s)\boldsymbol{\Gamma}\boldsymbol{g}(\gamma,\eta)]\}\beta(\tau)x(t)+\boldsymbol{n}(t,\tau)\\&=\{\boldsymbol{\alpha}_r(\theta_d)\otimes[\boldsymbol{\Phi}(\theta_d)\boldsymbol{g}(\gamma,\eta)]+\mathrm{e}^{\mathrm{j}\delta}\boldsymbol{\alpha}_r(\theta_s)\otimes[\boldsymbol{\Phi}(\theta_s)\boldsymbol{\Gamma}\boldsymbol{g}(\gamma,\eta)]\}\\&\quad\beta(\tau)[\boldsymbol{\alpha}_t(\theta_d)+\mathrm{e}^{\mathrm{j}\delta}\rho_h\boldsymbol{\alpha}_t(\theta_s)]^{\mathrm{T}}\boldsymbol{\varphi}(t)+\boldsymbol{n}(t,\tau)\in\mathbb{C}^{6M\times1}\end{aligned} \tag{6.5}$$

其中，$\beta(\tau)=\exp(\mathrm{j}2\pi f_d\tau)$ 为单个目标不同脉冲之间的复反射系数，是一个未知确定复常数，假设它服从 Swelling 2 型分布。$\boldsymbol{\Gamma}=\mathrm{diag}(\rho_h,\rho_v)$，其中 ρ_h 在式(6.3)中定义，另一个垂直极化的反射系数可用式(6.6)计算得到[264]：

$$\rho_v=\frac{\varepsilon\sin\theta_d-\sqrt{\varepsilon-\cos^2\theta_d}}{\varepsilon\sin\theta_d+\sqrt{\varepsilon-\cos^2\theta_d}} \tag{6.6}$$

其中：

$$\boldsymbol{\Phi}(\theta,\varphi)\overset{\mathrm{def}}{=}\begin{bmatrix}\cos\theta\cos\varphi & -\sin\varphi\\ \cos\theta\sin\varphi & \cos\varphi\\ -\sin\theta & 0\\ -\sin\varphi & -\cos\theta\cos\varphi\\ \cos\varphi & -\cos\theta\sin\varphi\\ 0 & \sin\theta\end{bmatrix} \tag{6.7}$$

$$\boldsymbol{g}(\gamma,\eta)\overset{\mathrm{def}}{=}\begin{bmatrix}\sin\gamma\,\mathrm{e}^{\mathrm{j}\eta}\\ \cos\gamma\end{bmatrix} \tag{6.8}$$

其中，θ 可以用 θ_d 或者 θ_s 表示，γ 为极化辅角，η 为极化相位差，φ 为方位角，这里假设已经估计得到，不失一般性，设置 $\varphi=90°$，为了方便，下面将 $\boldsymbol{\Phi}(\theta,\varphi)$ 记为 $\boldsymbol{\Phi}(\theta)$。

利用发射信号矢量 $\boldsymbol{\varphi}(t)$ 对上式做匹配滤波可得

$$\begin{aligned}\boldsymbol{Z} &= \int \boldsymbol{z}(t,\tau)\boldsymbol{\varphi}(t)^{\mathrm{H}}\mathrm{d}t \\ &= \{\boldsymbol{\alpha}_r(\theta_d)\otimes[\boldsymbol{\Phi}(\theta_d)\boldsymbol{g}(\gamma,\eta)] + \mathrm{e}^{\mathrm{j}\delta}\boldsymbol{\alpha}_r(\theta_s)\otimes[\boldsymbol{\Phi}(\theta_s)\boldsymbol{\Gamma}\boldsymbol{g}(\gamma,\eta)]\} \\ &\quad \beta(\tau)\,[\boldsymbol{\alpha}_t(\theta_d) + \mathrm{e}^{\mathrm{j}\delta}\rho_h\boldsymbol{\alpha}_t(\theta_s)]^{\mathrm{T}} + N(\tau) \in \mathbb{C}^{6M\times M}\end{aligned} \tag{6.9}$$

对上述数据进行矢量化操作可得

$$\begin{aligned}\boldsymbol{x}(\tau) &= \mathrm{vec}(\boldsymbol{Z}) = [\boldsymbol{\alpha}_t(\theta_d) + \mathrm{e}^{\mathrm{j}\delta}\rho_h\boldsymbol{\alpha}_t(\theta_s)]\otimes \\ &\quad \{\boldsymbol{\alpha}_r(\theta_d)\otimes[\boldsymbol{\Phi}(\theta_d)\boldsymbol{g}(\gamma,\eta)] + \mathrm{e}^{\mathrm{j}\delta}\boldsymbol{\alpha}_r(\theta_s)\otimes[\boldsymbol{\Phi}(\theta_s)\boldsymbol{\Gamma}\boldsymbol{g}(\gamma,\eta)]\}\beta(\tau) + \\ &\quad \mathrm{vec}[\boldsymbol{N}(\tau)] \in \mathbb{C}^{6M^2\times 1}\end{aligned} \tag{6.10}$$

假设原始噪声是零均值高斯随机过程，则有 $E[\boldsymbol{n}(t_1,\tau)\boldsymbol{n}(t_2,\tau)^{\mathrm{H}}] = \sigma^2\boldsymbol{I}_{6M}\omega(t_1-t_2)$，其中 ω 为冲激函数，σ^2 为噪声功率。则匹配滤波后噪声等于：

$$\bar{\boldsymbol{n}}(\tau) = \mathrm{vec}[\boldsymbol{N}(\tau)] = \mathrm{vec}\int_0^{T_p}\boldsymbol{n}(t,\tau)\boldsymbol{\varphi}(t)^{\mathrm{H}}\mathrm{d}t \tag{6.11}$$

下面通过计算噪声协方差矩阵来考察匹配滤波之后噪声的特性。

$$\begin{aligned}\boldsymbol{R}_n &= E\{\mathrm{vec}[\boldsymbol{N}(\tau)]\,\mathrm{vec}[\boldsymbol{N}(\tau)]^{\mathrm{H}}\} \\ &= E\int_0^{T_p}\int_0^{T_p}\boldsymbol{\varphi}(t_1)^*\otimes\boldsymbol{n}(t_1,\tau)[\boldsymbol{\varphi}^{\mathrm{T}}(t_2)\otimes\boldsymbol{n}(t_2,\tau)^{\mathrm{H}}]\,\mathrm{d}t_1\mathrm{d}t_2 \\ &= \int_0^{T_p}\int_0^{T_p}\boldsymbol{\varphi}(t_1)^*\boldsymbol{\varphi}^{\mathrm{T}}(t_2)\otimes E[\boldsymbol{n}(t_1,\tau)\boldsymbol{n}(t_2,\tau)^{\mathrm{H}}]\,\mathrm{d}t_1\mathrm{d}t_2 \\ &= \int_0^{T_p}\int_0^{T_p}\boldsymbol{\varphi}(t_1)^*\boldsymbol{\varphi}^{\mathrm{T}}(t_2)\otimes\sigma^2\boldsymbol{I}_{6M}\omega(t_1-t_2)\,\mathrm{d}t_1\mathrm{d}t_2 \\ &= \sigma^2\boldsymbol{I}_M\otimes\boldsymbol{I}_{6M} = \sigma^2\boldsymbol{I}_{6M^2}\end{aligned} \tag{6.12}$$

由式(6.12)可以看出，原始白噪声经过匹配滤波、矢量化操作之后，噪声的协方差矩阵仍然是一个单位阵，则噪声仍然是白噪声。

§6.3　两种参数估计方法

§6.3.1　信号模型的变形

米波极化 MIMO 雷达的测高信号接收模型与米波 MIMO 雷达的测高信号模型相比，其需要考虑的因素不仅有发射与反射之间相互渗透的问题，而且需要考虑极化信息，整个阵列的接收数据相对复杂，因此需要对上节的信号模型进行变形、归类，使其适用于超分辨算法。

将式(6.10)重写成如下形式：

$$\begin{aligned}\boldsymbol{x}(\tau)=&[\boldsymbol{\alpha}_t(\theta_d)+\mathrm{e}^{\mathrm{j}\delta}\rho_h\boldsymbol{\alpha}_t(\theta_s)]\otimes\\&\{\boldsymbol{\alpha}_r(\theta_d)\otimes[\boldsymbol{\Phi}(\theta_d)\boldsymbol{g}(\gamma,\eta)]+\mathrm{e}^{\mathrm{j}\delta}\boldsymbol{\alpha}_r(\theta_s)\otimes[\boldsymbol{\Phi}(\theta_s)\boldsymbol{\Gamma}\boldsymbol{g}(\gamma,\eta)]\}\otimes\\&\beta(\tau)+\bar{\boldsymbol{n}}(\tau)\in\mathbb{C}^{6M^2\times1}\end{aligned}\tag{6.13}$$

Kronecker 积满足结合率与分配率，可将式(6.13)转化为

$$\begin{aligned}\boldsymbol{x}(\tau)=&\Big[\boldsymbol{\alpha}_t(\theta_d)\otimes\boldsymbol{\alpha}_r(\theta_d)\otimes[\boldsymbol{\Phi}(\theta_d)\boldsymbol{g}(\gamma,\eta)]+\mathrm{e}^{\mathrm{j}\delta}\rho_h\boldsymbol{\alpha}_t(\theta_s)\otimes\boldsymbol{\alpha}_r(\theta_d)\otimes\\&[\boldsymbol{\Phi}(\theta_d)\boldsymbol{g}(\gamma,\eta)]+\boldsymbol{\alpha}_t(\theta_d)\otimes\mathrm{e}^{\mathrm{j}\delta}\boldsymbol{\alpha}_r(\theta_s)\otimes[\boldsymbol{\Phi}(\theta_s)\boldsymbol{\Gamma}\boldsymbol{g}(\gamma,\eta)]+\\&\mathrm{e}^{\mathrm{j}\delta}\rho_h\boldsymbol{\alpha}_t(\theta_s)\otimes\mathrm{e}^{\mathrm{j}\delta}\boldsymbol{\alpha}_r(\theta_s)\otimes[\boldsymbol{\Phi}(\theta_s)\boldsymbol{\Gamma}\boldsymbol{g}(\gamma,\eta)]\Big]\otimes\\&\beta(\tau)+\bar{\boldsymbol{n}}(\tau)\in\mathbb{C}^{6M^2\times1}\end{aligned}\tag{6.14}$$

式(6.14)可看作导向矢量合成的入射信号接收模型，且目标个数为1。将导向矢量归为一类、反射系数与波程差归为一类、极化信息归为一类，可将式(6.14)继续变形可得

$$\boldsymbol{x}(\tau)=\begin{bmatrix}\boldsymbol{\alpha}_t(\theta_d)\otimes\boldsymbol{\alpha}_r(\theta_d)\otimes\boldsymbol{\Phi}(\theta_d)\\\boldsymbol{\alpha}_t(\theta_d)\otimes\boldsymbol{\alpha}_r(\theta_s)\otimes\boldsymbol{\Phi}(\theta_s)\\\boldsymbol{\alpha}_t(\theta_s)\otimes\boldsymbol{\alpha}_r(\theta_d)\otimes\boldsymbol{\Phi}(\theta_d)\\\boldsymbol{\alpha}_t(\theta_s)\otimes\boldsymbol{\alpha}_r(\theta_s)\otimes\boldsymbol{\Phi}(\theta_s)\end{bmatrix}^{\mathrm{T}}\begin{bmatrix}\boldsymbol{I}_2\\\mathrm{e}^{\mathrm{j}\delta}\boldsymbol{\Gamma}\\\mathrm{e}^{\mathrm{j}\delta}\rho_h\boldsymbol{I}_2\\\mathrm{e}^{\mathrm{j}2\delta}\rho_h\boldsymbol{\Gamma}\end{bmatrix}\boldsymbol{g}(\gamma,\eta)\beta(\tau)+\bar{\boldsymbol{n}}(\tau)\tag{6.15}$$

因为反射系数 ρ_h,ρ_v 与波程差 δ 均为 θ_d,θ_s 的函数，因此为了方便可以做如下定义：

$$\boldsymbol{A}_{\mathrm{gmusic}}(\theta_d,\theta_s)=\begin{bmatrix}\boldsymbol{\alpha}_t(\theta_d)\otimes\boldsymbol{\alpha}_r(\theta_d)\otimes\boldsymbol{\Phi}(\theta_d)\\\boldsymbol{\alpha}_t(\theta_d)\otimes\boldsymbol{\alpha}_r(\theta_s)\otimes\boldsymbol{\Phi}(\theta_s)\\\boldsymbol{\alpha}_t(\theta_s)\otimes\boldsymbol{\alpha}_r(\theta_d)\otimes\boldsymbol{\Phi}(\theta_d)\\\boldsymbol{\alpha}_t(\theta_s)\otimes\boldsymbol{\alpha}_r(\theta_s)\otimes\boldsymbol{\Phi}(\theta_s)\end{bmatrix}^{\mathrm{T}}\in\mathbb{C}^{6M^2\times8}\tag{6.16}$$

$$\boldsymbol{A}_f(\rho_h,\rho_v)=\begin{bmatrix}\boldsymbol{I}_2\\\mathrm{e}^{\mathrm{j}\delta}\boldsymbol{\Gamma}\\\mathrm{e}^{\mathrm{j}\delta}\rho_h\boldsymbol{I}_2\\\mathrm{e}^{\mathrm{j}2\delta}\rho_h\boldsymbol{\Gamma}\end{bmatrix}\in\mathbb{C}^{8\times2}\tag{6.17}$$

$$\boldsymbol{A}_{\mathrm{music}}(\theta_d,\theta_s,\rho_h,\rho_v)=\boldsymbol{A}_{\mathrm{gmusic}}(\theta_d,\theta_s)\boldsymbol{A}_f(\rho_h,\rho_v)\in\mathbb{C}^{6M^2\times2}\tag{6.18}$$

根据上面三个定义，式(6.13)可简写成

$$\begin{aligned}\boldsymbol{x}(\tau)&=\boldsymbol{A}_{\mathrm{music}}(\theta_d,\theta_s,\rho_h,\rho_v)\boldsymbol{g}(\gamma,\eta)\beta(\tau)+\bar{n}(\tau)\\&=\boldsymbol{A}_{\mathrm{gmusic}}(\theta_d,\theta_s)\boldsymbol{A}_f(\rho_h,\rho_v)\boldsymbol{g}(\gamma,\eta)\beta(\tau)+\bar{\boldsymbol{n}}(\tau)\end{aligned}\tag{6.19}$$

§ 6.3.2　广义 MUSIC

对于式(6.19)，可将 $\boldsymbol{A}_f(\rho_h,\rho_v)\boldsymbol{g}(\gamma,\eta)\beta(\tau)$ 看成是 8 个相干入射信号源，即定义新的信号源 $\bar{\beta}(\rho_h,\rho_v,\gamma,\eta,\tau)=\boldsymbol{A}_f(\rho_h,\rho_v)\boldsymbol{g}(\gamma,\eta)\beta(\tau)$，注意到其不同的角度只有两个，即直达角和反射角 θ_d,θ_s，即在谱峰搜索的超分辨算法中只能得到两个峰值，则可认为在划分信号子空间时，入射信号源为 2。则接收信号模型变为

$$\boldsymbol{x}(\tau)=\boldsymbol{A}_{\text{gmusic}}(\theta_d,\theta_s)\bar{\beta}(\rho_h,\rho_v,\gamma,\eta,\tau)+\bar{\boldsymbol{n}}(\tau) \tag{6.20}$$

式(6.20)可利用标准的广义 MUSIC 算法来得到两个相干目标 θ_d,θ_s 的角度估计值。首先对式(6.20)的接收数据协方差 $\boldsymbol{R}=E\{\boldsymbol{x}(\tau)*\boldsymbol{x}(\tau)^{\mathrm{H}}\}$ 进行特征值分解，把两个大特征值对应的特征向量构成的空间称为信号子空间 $\boldsymbol{E}_s$，其余 $6M^2\times(6M^2-2)$ 维特征向量构成噪声子空间 $\boldsymbol{E}_n$，再根据文献[265]的推导可以得到米波极化 MIMO 雷达的广义 MUSIC 空间谱为

$$P(\theta_d,\theta_s)=\frac{\det[\boldsymbol{A}_{\text{gmusic}}{}^{\mathrm{H}}(\theta_d,\theta_s)\boldsymbol{A}_{\text{gmusic}}(\theta_d,\theta_s)]}{\det[\boldsymbol{A}_{\text{gmusic}}{}^{\mathrm{H}}(\theta_d,\theta_s)\boldsymbol{E}_n\boldsymbol{E}_n{}^{\mathrm{H}}\boldsymbol{A}_{\text{gmusic}}(\theta_d,\theta_s)]} \tag{6.21}$$

式(6.21)同样涉及二维搜索，计算量较大，可利用直达波 θ_d 与反射波 θ_s，几何关系式为

$$\theta_s=-\arctan\left[\tan(\theta_d)+\frac{2h_a}{R}\right]\approx-\theta_d \tag{6.22}$$

因此可以利用上式将二维搜索变为一维搜索，大大降低计算量。从上面的解法可知广义 MUSIC 算法的优点是其与反射系数无关，对阵地具有较强的鲁棒性，缺点是其无法估计极化参数。下面提出导向矢量合成降维 MUSIC 算法，该算法可同时估计直达波角度 θ_d 和极化参数 γ,η。

§ 6.3.3　导向矢量合成降维 MUSIC

从式(6.19)可以发现，信号模型经过变形处理后，$\beta(\tau)$ 为经过匹配滤波后的接收信号，则 $\boldsymbol{A}_{\text{music}}(\theta_d,\theta_s,\rho_h,\rho_v)\boldsymbol{g}(\gamma,\eta)$ 可以视为信号导向矢量。不难发现，$\boldsymbol{A}_{\text{music}}(\theta_d,\theta_s,\rho_h,\rho_v)\boldsymbol{g}(\gamma,\eta)$ 为 $6M^2\times1$ 维，其秩为 1。由此可以观察到，经过变形后的信号模型导向矢量成功地规避了反射波的影响，即将反射波形成的导向矢量合成到直达波导向矢量，故该算法的前缀冠上导向矢量合成技术。秩为 1 即是只有一个入射信号源的阵列接收信号模型，因此，许多常规的超分辨 DOA 估计算法可直接应用于此信号模型，无需解相干处理。本书将常规 MUSIC 算法直接应用在信号模型上，还进行了降维处理。下面给出具体

推导：

对式(6.19)的极化 MIMO 雷达的接收数据的接收信号协方差进行特征值分解，对特征向量进行划分，把唯一的一个大特征值对应的特征向量构成信号子空间$\bar{\boldsymbol{E}}_s$，其余 $6M^2\times(6M^2-1)$ 维特征向量构成噪声子空间$\bar{\boldsymbol{E}}_n$，再根据常规 MUSIC 算法可以得到米波极化 MIMO 雷达空间谱如下：

$$P(\theta_d,\theta_s,\rho_h,\rho_v,\gamma,\eta)=\frac{1}{[\boldsymbol{A}_{\text{music}}(\theta_d,\theta_s,\rho_h,\rho_v)\mathrm{g}(\gamma,\eta)]^{\mathrm{H}}\bar{\boldsymbol{E}}_n\bar{\boldsymbol{E}}_n^{\mathrm{H}}\boldsymbol{A}_{\text{music}}(\theta_d,\theta_s,\rho_h,\rho_v)\boldsymbol{g}(\gamma,\eta)} \tag{6.23}$$

其中，式(6.23)中含有六个未知数，需要进行六维搜索处理，这个计算量并不适合实际工程应用，因此需要降维处理。将降维分为两个阶段进行：第一阶段将极化信息与 DOA 信息解耦合来降维；第二阶段利用直达波与反射波的关系，以及反射系数与直达波的关系来降维。

下面来看第一阶段的降维处理，先定义 MUSIC 代价函数：

$$V=[\boldsymbol{A}_{\text{music}}(\theta_d,\theta_s,\rho_h,\rho_v)\boldsymbol{g}(\gamma,\eta)]^{\mathrm{H}}\bar{\boldsymbol{E}}_n\bar{\boldsymbol{E}}_n^{\mathrm{H}}\boldsymbol{A}_{\text{music}}(\theta_d,\theta_s,\rho_h,\rho_v)\boldsymbol{g}(\gamma,\eta) \tag{6.24}$$

不难发现 $\boldsymbol{g}(\gamma,\eta)^{\mathrm{H}}\boldsymbol{g}(\gamma,\eta)=1$，因此可将代价函数进行变形为

$$V=\frac{[\boldsymbol{A}_{\text{music}}(\theta_d,\theta_s,\rho_h,\rho_v)\boldsymbol{g}(\gamma,\eta)]^{\mathrm{H}}\bar{\boldsymbol{E}}_n\bar{\boldsymbol{E}}_n^{\mathrm{H}}\boldsymbol{A}_{\text{music}}(\theta_d,\theta_s,\rho_h,\rho_v)\boldsymbol{g}(\gamma,\eta)}{\boldsymbol{g}(\gamma,\eta)^{\mathrm{H}}\boldsymbol{g}(\gamma,\eta)} \tag{6.25}$$

通过观察式(6.25)可以发现，其符合瑞利商求最大或者最小值的准则，因此有下式成立：

$$\lambda_{\min}(\theta_d,\theta_s)=\min_{g(\gamma,\eta)\neq 0}(V) \tag{6.26}$$

其中，$\lambda_{\min}(\theta_d,\theta_s)$表示矩阵$\boldsymbol{A}_{\text{music}}(\theta_d,\theta_s,\rho_h,\rho_v)^{\mathrm{H}}\bar{\boldsymbol{E}}_n\bar{\boldsymbol{E}}_n^{\mathrm{H}}\boldsymbol{A}_{\text{music}}(\theta_d,\theta_s,\rho_h,\rho_v)$特征分解得到的最小特征值。则可通过极化信息解耦得到四维搜索的空间谱来估计目标的直达波和反射波，即

$$f_{\text{4D-MUSIC}}=\arg\max_{\theta_d,\theta_s,\rho_h,\rho_v}[1/\lambda_{\min}(\theta_d,\theta_s,\rho_h,\rho_v)] \tag{6.27}$$

极化矢量的估计值$\hat{\boldsymbol{g}}(\gamma,\eta)$等于最小特征值对应的特征矢量，因此极化参数可以通过式(6.28)得到：

$$\left.\begin{aligned}\hat{\gamma}&=\arctan\frac{[\hat{\boldsymbol{g}}(\gamma,\eta)]_1}{[\hat{\boldsymbol{g}}(\gamma,\eta)]_2}\\ \hat{\eta}&=\angle\frac{[\hat{\boldsymbol{g}}(\gamma,\eta)]_1}{[\hat{\boldsymbol{g}}(\gamma,\eta)]_2}\end{aligned}\right\} \tag{6.28}$$

从式(6.27)可看出极化信息与DOA信息已经解耦合,将搜索降低了两个维度,但四维搜索的计算量依然不可接受,下面进行第二阶段的降维处理。

基于式(6.22)的直达波和反射波的几何关系,可将式(6.27)降至三维搜索。三维角度搜索仍然不可接受,需要继续降维。从式(6.3)和式(6.6)可知,反射系数 ρ_h、ρ_v 由直达波和反射波 θ_d、相对介电常数 ε_r 和表面物质传导率 σ_e 决定。在不同的阵地场景下,如水面、陆地、植被等场景不尽相同,幸运的是已有前辈学者们测定总结了不同场景下的具体值,见表 6.1。根据表 6.1,相对介电常数 ε_r 和表面物质传导率 σ_e 已知,式(6.27)可降维至一维搜索,此时就完成了 MUSIC 的降维处理,故称该算法为导向矢量合成降维 MUSIC 算法。

表 6.1　不同地形下的相对介电常数 ε_r 和表面物质传导率 σ_e[264]

序　号	介　质	相对介电常数	表面物质传导率
1	良好的土壤(湿土)	25	0.02
2	一般土壤	15	0.005
3	贫瘠的土壤(干土)	3	0.001
4	雪、冰	3	0.001
5	淡水	81	0.7
6	盐水	75	0.5

需要注意的是,在复杂的阵地场景下,相对介电常数 ε_r 和表面物质传导率 σ_e 不能够精确测定,是有一定误差的,此时就需要三维搜索,可利用一种交替搜索的方法来解决上述问题[266],但这不是本章核心内容,故不做展开。

§6.3.4　计算复杂度

本章的广义 MUSIC 计算复杂度主要取决于:接收信号协方差矩阵的计算,该部分需要 $O\{(6M^2)^2L\}$ 复乘运算;特征值分解,该部分需要 $O\{(6M^2)^3\}$ 复乘运算;空间谱搜索,该部分需要 $O\{n[64(6M^2)+8^3+8\otimes 2\otimes(6M^2)+32+8^3]\}$ 复乘运算。因此,广义 MUSIC 算法总共需要的复乘运算为:$O\{(6M^2)^2L+(6M^2)^3+n[64(6M^2)+8^3+8\otimes 2\otimes(6M^2)+32+8^3]\}$,其中 n 为一维角度搜索设置的角度搜索间隔。

利用相对介电常数 ε_r 和表面物质传导率 σ_e 已知情况下的导向矢量合成降维,常规 MUSIC 计算量也分为三个部分:接收信号协方差矩阵计算、特征值分解、空间谱搜索。其计算量分别为:$O\{(6M^2)^2L\}$,$O((6M^2)^3)$,

$O\{n[2(6M^2)+12]\}$。因此，本章导向矢量合成降维 MUSIC 计算复杂度为 $O\{(6M^2)^2L+(6M^2)^3+n[2(6M^2)+12]\}$。下面给出计算量随阵元数的变化曲线，如图 6.2 所示，其中角度搜索间隔为 $n=1\ 000$ 个点，快拍数为 $L=10$。可以看出导向矢量合成算法的计算量比广义 MUSIC 算法的计算量少。但是在反射系数未知的情况下，导向矢量合成 MUSIC 涉及三维搜索，计算量是很大的。

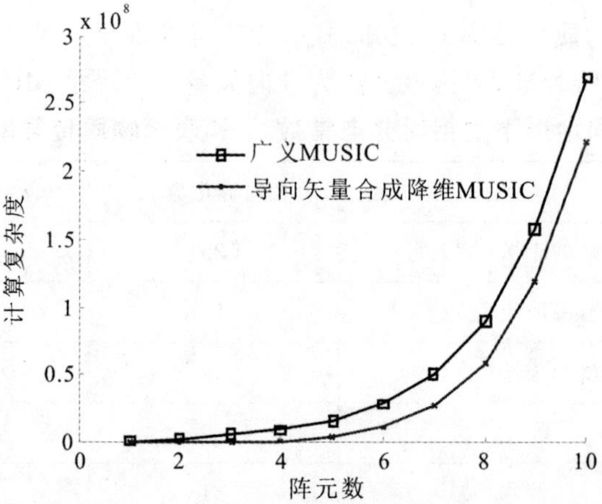

图 6.2　广义 MUSIC 和导向矢量合成降维 MUSIC 计算量对比图

§ 6.3.5　CRB

假设反射系数中相对介电常数 ε_r 和表面物质传导率 σ_e 已知，直达波和反射波的几何关系固定。则待估计的未知参数有三个，即 (θ_d,γ,η)。其 FIM 可用如下矩阵表示：

$$\boldsymbol{J}=\begin{bmatrix} J_{\theta_d\theta_d} & J_{\theta_d\gamma} & J_{\theta_d\eta} \\ J_{\gamma\theta_d} & J_{\gamma\gamma} & J_{\gamma\eta} \\ J_{\eta\theta_d} & J_{\eta\gamma} & J_{\eta\eta} \end{bmatrix}_{3\times 3} \tag{6.29}$$

L 次快拍的 FIM 的第 (i,j) 个元素等于：

$$\boldsymbol{J}(i,j)=L\,\mathrm{Tr}\left\{\boldsymbol{R}^{-1}\frac{\partial\boldsymbol{R}}{\partial\alpha_i}\boldsymbol{R}^{-1}\frac{\partial\boldsymbol{R}}{\partial\alpha_j}\right\} \tag{6.30}$$

其中，$\boldsymbol{R}$ 为接收数据协方差矩阵，数据中只有一个目标个数，$\boldsymbol{R}$ 可展开为

$$\boldsymbol{R}=\sigma_s^2\boldsymbol{A}_{\mathrm{music}}(\theta_d,\theta_s,\rho_h,\rho_v)\boldsymbol{g}(\gamma,\eta)\left[\boldsymbol{A}_{\mathrm{music}}(\theta_d,\theta_s,\rho_h,\rho_v)\boldsymbol{g}(\gamma,\eta)\right]^{\mathrm{H}}+\sigma_n^2\boldsymbol{I}_{6M^2} \tag{6.31}$$

其中 σ_s^2 是信号的功率。下面要注意符号的缩写。定义：

$$\boldsymbol{\alpha}(\theta_d,\gamma,\eta)\overset{\text{def}}{=}\boldsymbol{A}_{\text{music}}(\theta_d,\theta_s,\rho_h,\rho_v)g(\gamma,\eta)=\boldsymbol{A}_{\text{gmusic}}(\theta_d,\theta_s)\boldsymbol{A}_f(\rho_h,\rho_v)g(\gamma,\eta)$$

其中，ρ_h,ρ_v 是 θ_d 的函数，故在下面的推导中 $\boldsymbol{A}_f(\rho_h,\rho_v)$ 记为 $\boldsymbol{A}_f(\theta_d)$。则协方差矩阵 $\boldsymbol{R}$ 对目标三维参数 (θ_d,γ,η) 的偏导分别为

$$\frac{\partial \boldsymbol{R}}{\partial \theta_d}=\frac{\sigma_s^2\partial \boldsymbol{\alpha}\boldsymbol{\alpha}^{\text{H}}}{\partial \theta_d}=\sigma_s^2\frac{\partial \boldsymbol{\alpha}}{\partial \theta_d}\boldsymbol{\alpha}^{\text{H}}+\sigma_s^2\boldsymbol{\alpha}\frac{\partial \boldsymbol{\alpha}^{\text{H}}}{\partial \theta_d} \tag{6.32}$$

$$\frac{\partial \boldsymbol{R}}{\partial \gamma}=\frac{\sigma_s^2\partial \boldsymbol{\alpha}\boldsymbol{\alpha}^{\text{H}}}{\partial \gamma}=\sigma_s^2\frac{\partial \boldsymbol{\alpha}}{\partial \gamma}\boldsymbol{\alpha}^{\text{H}}+\sigma_s^2 a\frac{\partial \boldsymbol{\alpha}^{\text{H}}}{\partial \gamma} \tag{6.33}$$

$$\frac{\partial \boldsymbol{R}}{\partial \eta}=\frac{\sigma_s^2\partial \boldsymbol{\alpha}\boldsymbol{\alpha}^{\text{H}}}{\partial \eta}=\sigma_s^2\frac{\partial \boldsymbol{\alpha}}{\partial \eta}\boldsymbol{\alpha}^{\text{H}}+\sigma_s^2\boldsymbol{\alpha}\frac{\partial \boldsymbol{\alpha}^{\text{H}}}{\partial \eta} \tag{6.34}$$

下面给出 $\frac{\partial \boldsymbol{\alpha}^{\text{H}}}{\partial \theta_d}$、$\frac{\partial \boldsymbol{\alpha}^{\text{H}}}{\partial \gamma}$ 和 $\frac{\partial \boldsymbol{\alpha}^{\text{H}}}{\partial \eta}$ 的推导结果：

$$\bar{\boldsymbol{\alpha}}_{\theta_d}\overset{\text{def}}{=}\frac{\partial \boldsymbol{\alpha}}{\partial \theta_d}=\frac{\partial \boldsymbol{A}_{\text{gmusic}}(\theta_d,\theta_s)}{\partial \theta_d}\boldsymbol{A}_f(\theta_d)\boldsymbol{g}(\gamma,\eta)+\boldsymbol{A}_{\text{gmusic}}(\theta_d,\theta_s)\frac{\partial \boldsymbol{A}_{\text{f}}(\theta_d)}{\partial \theta_d}\boldsymbol{g}(\gamma,\eta) \tag{6.35}$$

将 $\boldsymbol{A}_{\text{gmusic}}(\theta_d,\theta_s)$ 拆分成四个模块，即 $\boldsymbol{A}_{\text{gmusic}}(\theta_d,\theta_s)=[\boldsymbol{A}_1\quad \boldsymbol{A}_2\quad \boldsymbol{A}_3\quad \boldsymbol{A}_4]$，则

$$\left.\begin{aligned}\boldsymbol{A}_1&=\boldsymbol{\alpha}_{\text{t}}(\theta_d)\otimes\boldsymbol{\alpha}_r(\theta_d)\otimes\boldsymbol{\Phi}(\theta_d)\\ \boldsymbol{A}_2&=\boldsymbol{\alpha}_{\text{t}}(\theta_d)\otimes\boldsymbol{\alpha}_r(\theta_s)\otimes\boldsymbol{\Phi}(\theta_s)\\ \boldsymbol{A}_3&=\boldsymbol{\alpha}_{\text{t}}(\theta_s)\otimes\boldsymbol{\alpha}_r(\theta_d)\otimes\boldsymbol{\Phi}(\theta_d)\\ \boldsymbol{A}_4&=\boldsymbol{a}_{\text{t}}(\theta_s)\otimes\boldsymbol{\alpha}_r(\theta_s)\otimes\boldsymbol{\Phi}(\theta_s)\end{aligned}\right\} \tag{6.36}$$

可以得到

$$\frac{\partial \boldsymbol{A}_{\text{gmusic}}(\theta_d,\theta_s)}{\partial \theta_d}=\begin{bmatrix}\frac{\partial \boldsymbol{A}_1}{\partial \theta_d} & \frac{\partial \boldsymbol{A}_2}{\partial \theta_d} & \frac{\partial \boldsymbol{A}_3}{\partial \theta_d} & \frac{\partial \boldsymbol{A}_4}{\partial \theta_d}\end{bmatrix} \tag{6.37}$$

则有如下几个公式：

$$\begin{aligned}\frac{\partial \boldsymbol{A}_1}{\partial \theta_d}=&[\boldsymbol{c}_1\odot\boldsymbol{\alpha}_t(\theta_d)]\otimes\boldsymbol{\alpha}_r(\theta_d)\otimes\boldsymbol{\Phi}(\theta_d)+\boldsymbol{\alpha}_t(\theta_d)\otimes[\boldsymbol{c}_1\odot\boldsymbol{\alpha}_r(\theta_d)]\otimes\boldsymbol{\Phi}(\theta_d)+\\&\boldsymbol{\alpha}_t(\theta_d)\otimes\boldsymbol{\alpha}_r(\theta_d)\otimes[\boldsymbol{C}_2\odot\boldsymbol{\Phi}(\theta_d)]\end{aligned} \tag{6.38}$$

$$\begin{aligned}\frac{\partial \boldsymbol{A}_2}{\partial \theta_d}=&[\boldsymbol{c}_1\odot\boldsymbol{\alpha}_t(\theta_d)]\otimes\boldsymbol{\alpha}_r(\theta_s)\otimes\boldsymbol{\Phi}(\theta_s)+\boldsymbol{\alpha}_t(\theta_d)\otimes[\boldsymbol{c}_3\odot\boldsymbol{\alpha}_r(\theta_s)]\otimes\boldsymbol{\Phi}(\theta_s)+\\&\boldsymbol{\alpha}_t(\theta_d)\otimes\boldsymbol{\alpha}_r(\theta_s)\otimes[\boldsymbol{C}_4\odot\boldsymbol{\Phi}(\theta_s)]\end{aligned} \tag{6.39}$$

$$\begin{aligned}\frac{\partial \boldsymbol{A}_3}{\partial \theta_d}=&[\boldsymbol{c}_3\odot\boldsymbol{\alpha}_t(\theta_s)]\otimes\boldsymbol{\alpha}_r(\theta_d)\otimes\boldsymbol{\Phi}(\theta_d)+\boldsymbol{\alpha}_t(\theta_s)\otimes[\boldsymbol{c}_1\odot\boldsymbol{\alpha}_r(\theta_d)]\otimes\boldsymbol{\Phi}(\theta_d)+\\&\boldsymbol{\alpha}_t(\theta_s)\otimes\boldsymbol{\alpha}_r(\theta_d)\otimes[\boldsymbol{C}_2\odot\boldsymbol{\Phi}(\theta_d)]\end{aligned} \tag{6.40}$$

$$\frac{\partial \boldsymbol{A}_4}{\partial \theta_d}=[\boldsymbol{c}_3\odot\boldsymbol{\alpha}_t(\theta_s)]\otimes\boldsymbol{\alpha}_r(\theta_s)\otimes\boldsymbol{\Phi}(\theta_s)+\boldsymbol{\alpha}_t(\theta_s)\otimes[\boldsymbol{c}_3\odot\boldsymbol{\alpha}_r(\theta_s)]\otimes\boldsymbol{\Phi}(\theta_s)+$$
$$\boldsymbol{\alpha}_t(\theta_s)\otimes\boldsymbol{\alpha}_r(\theta_s)\otimes[\boldsymbol{C}_4\odot\boldsymbol{\Phi}(\theta_s)] \tag{6.41}$$

上式中：

$$\boldsymbol{c}_1=-j\frac{2\pi}{\lambda}d\cos\theta_d\,[0\ 1\ \cdots\ (M-1)]^{\mathrm{T}} \tag{6.42}$$

$$\boldsymbol{C}_2=\begin{bmatrix}-\sin\theta_d\cos\phi & -\sin\theta_d\sin\phi & -\cos\theta_d & 0 & 0 & 0\\ 0 & 0 & 0 & \sin\theta_d\cos\phi & \sin\theta_d\sin\phi & -\cos\theta_d\end{bmatrix}^{\mathrm{T}} \tag{6.43}$$

$$\boldsymbol{c}_3=-\bar{\theta}_s j\frac{2\pi}{\lambda}d\cos\theta_s\,[0\ 1\ \cdots\ (M-1)]^{\mathrm{T}} \tag{6.44}$$

$$\boldsymbol{C}_4=\bar{\theta}_s\begin{bmatrix}-\sin\theta_s\cos\phi & -\sin\theta_s\sin\phi & -\cos\theta_s & 0 & 0 & 0\\ 0 & 0 & 0 & 0\sin\theta_s\cos\phi & \sin\theta_s\sin\phi & -\cos\theta_s\end{bmatrix}^{\mathrm{T}} \tag{6.45}$$

$$\bar{\theta}_s\overset{\mathrm{def}}{=}\frac{\partial\theta_s}{\partial\theta_d}=-\frac{1+\tan\theta_d{}^2}{1+\left(\tan\theta_d+\frac{2h_a}{R}\right)^2} \tag{6.46}$$

$$\bar{\boldsymbol{A}}_f(\theta_d)=\begin{bmatrix}\mathbf{O}_2\\ \mathrm{e}^{\mathrm{j}\delta}\frac{4\pi h_a}{\lambda}(1+\tan\theta_d{}^2)\boldsymbol{\Gamma}+\mathrm{e}^{\mathrm{j}\delta}diag(\bar{\rho}_h,\bar{\rho}_v)\\ \mathrm{e}^{\mathrm{j}\delta}\frac{4\pi h_a}{\lambda}(1+\tan\theta_d{}^2)\rho_h\boldsymbol{I}_2+\mathrm{e}^{\mathrm{j}\delta}\bar{\rho}_h\boldsymbol{I}_2\\ \mathrm{e}^{\mathrm{j}2\delta}\frac{8\pi h_a}{\lambda}(1+\tan\theta_d{}^2)\rho_h\boldsymbol{\Gamma}+\mathrm{e}^{\mathrm{j}2\delta}\bar{\rho}_h\boldsymbol{\Gamma}+\mathrm{e}^{\mathrm{j}2\delta}\rho_h\,\mathrm{diag}(\bar{\rho}_h,\bar{\rho}_v)\end{bmatrix} \tag{6.47}$$

$$\bar{\rho}_h=\frac{(\cos\theta_d-b)\left(\sin\theta_d+\sqrt{\varepsilon-\cos^2\theta_d}\right)-\left(\sin\theta_d-\sqrt{\varepsilon-\cos^2\theta_d}\right)(\cos\theta_d+b)}{\left(\sin\theta_d+\sqrt{\varepsilon-\cos^2\theta_d}\right)^2} \tag{6.48}$$

$$\bar{\rho}_v=\frac{(\varepsilon\cos\theta_d-b)\left(\varepsilon\sin\theta_d+\sqrt{\varepsilon-\cos^2\theta_d}\right)-\left(\varepsilon\sin\theta_d-\sqrt{\varepsilon-\cos^2\theta_d}\right)(\varepsilon\cos\theta_d+b)}{\left(\varepsilon\sin\theta_d+\sqrt{\varepsilon-\cos^2\theta_d}\right)^2} \tag{6.49}$$

$$b=-\sqrt{\frac{1}{\varepsilon-\cos^2\theta_d}}\cos\theta_d\sin\theta_d \tag{6.50}$$

$$\bar{\boldsymbol{\alpha}}_\gamma \stackrel{\text{def}}{=} \frac{\partial \boldsymbol{\alpha}}{\partial \gamma} = \boldsymbol{A}_{\text{music}}(\theta_d, \theta_s)c_5 \tag{6.51}$$

$$\bar{\boldsymbol{\alpha}}_\eta \stackrel{\text{def}}{=} \frac{\partial \boldsymbol{\alpha}}{\partial \eta} = \boldsymbol{A}_{\text{music}}(\theta_d, \theta_s)c_6 \tag{6.52}$$

其中：

$$\boldsymbol{c}_5 = [\cos\gamma e^{j\eta} \quad -\sin\gamma]^{\mathrm{T}} \tag{6.53}$$

$$\boldsymbol{c}_6 = [j\sin\gamma e^{j\eta} \quad 0]^{\mathrm{T}} \tag{6.54}$$

把式(6.32)和式(6.34)计算得到的结果代入式(6.35)，则 $\boldsymbol{J}_{\theta_d \eta}$ 等于：

$$\begin{aligned}
\boldsymbol{J}_{\theta_d \eta} &= L\,\mathrm{Tr}\{\boldsymbol{R}^{-1}(\sigma_s^2 \bar{\boldsymbol{a}}_{\theta_d}\boldsymbol{a}^{\mathrm{H}} + \sigma_s^2 \boldsymbol{a}\,\bar{\boldsymbol{a}}_{\theta_d}^{\mathrm{H}})\boldsymbol{R}^{-1}(\sigma_s^2 \bar{\boldsymbol{a}}_\eta \boldsymbol{a}^{\mathrm{H}} + \sigma_s^2 \boldsymbol{a}\,\bar{\boldsymbol{a}}_\eta^{\mathrm{H}})\} \\
&= L\sigma_s^2\sigma_s^2\,\mathrm{Tr}\{\boldsymbol{R}^{-1}(\bar{\boldsymbol{a}}_{\theta_d}\boldsymbol{a}^{\mathrm{H}} + \boldsymbol{a}\,\bar{\boldsymbol{a}}_{\theta_d}^{\mathrm{H}})\boldsymbol{R}^{-1}(\bar{\boldsymbol{a}}_\eta \boldsymbol{a}^{\mathrm{H}} + \boldsymbol{a}\,\bar{\boldsymbol{a}}_\eta^{\mathrm{H}})\} \\
&= L\sigma_s^2\sigma_s^2\,\mathrm{Tr}\Big\{\boldsymbol{R}^{-1}\bar{\boldsymbol{a}}_{\theta_d}\boldsymbol{a}^{\mathrm{H}}\boldsymbol{R}^{-1}\bar{\boldsymbol{a}}_\eta \boldsymbol{a}^{\mathrm{H}} + \boldsymbol{R}^{-1}\bar{\boldsymbol{a}}_{\theta_d}\boldsymbol{a}^{\mathrm{H}}\boldsymbol{R}^{-1}\boldsymbol{a}\,\bar{\boldsymbol{a}}_\eta^{\mathrm{H}} + \boldsymbol{R}^{-1}\boldsymbol{a}\,\bar{\boldsymbol{a}}_{\theta_d}^{\mathrm{H}}\boldsymbol{R}^{-1}\bar{\boldsymbol{a}}_\eta \boldsymbol{a}^{\mathrm{H}} + \\
&\qquad \boldsymbol{R}^{-1}\boldsymbol{a}\,\bar{\boldsymbol{a}}_{\theta_d}^{\mathrm{H}}\boldsymbol{R}^{-1}\boldsymbol{a}\,\bar{\boldsymbol{a}}_\eta^{\mathrm{H}}\Big\} \\
&= L\sigma_s^2\sigma_s^2\Big\{\boldsymbol{a}^{\mathrm{H}}\boldsymbol{R}^{-1}\bar{\boldsymbol{a}}_{\theta_d}\boldsymbol{a}^{\mathrm{H}}\boldsymbol{R}^{-1}\bar{\boldsymbol{a}}_\eta + \bar{\boldsymbol{a}}_\eta^{\mathrm{H}}\boldsymbol{R}^{-1}\bar{\boldsymbol{a}}_{\theta_d}\boldsymbol{a}^{\mathrm{H}}\boldsymbol{R}^{-1}\boldsymbol{a} + \boldsymbol{a}^{\mathrm{H}}\boldsymbol{R}^{-1}\boldsymbol{a}\,\bar{\boldsymbol{a}}_{\theta_d}^{\mathrm{H}}\boldsymbol{R}^{-1}\bar{\boldsymbol{a}}_\eta + \\
&\qquad \bar{\boldsymbol{a}}_\eta^{\mathrm{H}}\boldsymbol{R}^{-1}\boldsymbol{a}\,\bar{\boldsymbol{a}}_{\theta_d}^{\mathrm{H}}\boldsymbol{R}^{-1}\boldsymbol{a}\Big\} \\
&= 2L\sigma_s^2\sigma_s^2\,\mathrm{Re}\{\boldsymbol{a}^{\mathrm{H}}\boldsymbol{R}^{-1}\bar{\boldsymbol{a}}_\eta \boldsymbol{a}^{\mathrm{H}}\boldsymbol{R}^{-1}\bar{\boldsymbol{a}}_{\theta_d} + \boldsymbol{a}^{\mathrm{H}}\boldsymbol{R}^{-1}\boldsymbol{a}\,\bar{\boldsymbol{a}}_{\theta_d}^{\mathrm{H}}\boldsymbol{R}^{-1}\bar{\boldsymbol{a}}_\eta\}
\end{aligned} \tag{6.55}$$

类似 $\boldsymbol{J}_{\theta_d \eta}$ 的推导可得式(6.35)其他 8 个元素的表达式：

$$\boldsymbol{J}_{\eta\theta_d} = \boldsymbol{J}_{\theta_d \eta} \tag{6.56}$$

$$\boldsymbol{J}_{\theta_d\theta_d} = 2L\ (\sigma_s^2)^2\mathrm{Re}\{\boldsymbol{\alpha}^{\mathrm{H}}\boldsymbol{R}^{-1}\bar{\boldsymbol{\alpha}}_{\theta_d}\boldsymbol{\alpha}^{\mathrm{H}}\boldsymbol{R}^{-1}\bar{\boldsymbol{\alpha}}_{\theta_d} + \boldsymbol{\alpha}^{\mathrm{H}}\boldsymbol{R}^{-1}\boldsymbol{\alpha}\,\bar{\boldsymbol{\alpha}}_{\theta_d}^{\mathrm{H}}\boldsymbol{R}^{-1}\bar{\boldsymbol{\alpha}}_{\theta_d}\} \tag{6.57}$$

$$\boldsymbol{J}_{\gamma\theta_d} = \boldsymbol{J}_{\theta_d\gamma} = 2L\ (\sigma_s^2)^2\mathrm{Re}\{\boldsymbol{\alpha}^{\mathrm{H}}\boldsymbol{R}^{-1}\bar{\boldsymbol{\alpha}}_\gamma \boldsymbol{\alpha}^{\mathrm{H}}\boldsymbol{R}^{-1}\bar{\boldsymbol{\alpha}}_{\theta_d} + \boldsymbol{\alpha}^{\mathrm{H}}\boldsymbol{R}^{-1}\boldsymbol{\alpha}\,\bar{\boldsymbol{\alpha}}_{\theta_d}^{\mathrm{H}}\boldsymbol{R}^{-1}\bar{\boldsymbol{\alpha}}_\gamma\} \tag{6.58}$$

$$\boldsymbol{J}_{\gamma\gamma} = 2L\ (\sigma_s^2)^2\mathrm{Re}\{\boldsymbol{\alpha}^{\mathrm{H}}\boldsymbol{R}^{-1}\bar{\boldsymbol{\alpha}}_\gamma\boldsymbol{\alpha}^{\mathrm{H}}\boldsymbol{R}^{-1}\bar{\boldsymbol{\alpha}}_\gamma + \boldsymbol{\alpha}^{\mathrm{H}}\boldsymbol{R}^{-1}\boldsymbol{\alpha}\,\bar{\boldsymbol{\alpha}}_\gamma^{\mathrm{H}}\boldsymbol{R}^{-1}\bar{\boldsymbol{\alpha}}_\gamma\} \tag{6.59}$$

$$\boldsymbol{J}_{\eta\eta} = 2L\ (\sigma_s^2)^2\mathrm{Re}\{\boldsymbol{\alpha}^{\mathrm{H}}\boldsymbol{R}^{-1}\bar{\boldsymbol{\alpha}}_\eta\boldsymbol{\alpha}^{\mathrm{H}}\boldsymbol{R}^{-1}\bar{\boldsymbol{\alpha}}_\eta + \boldsymbol{\alpha}^{\mathrm{H}}\boldsymbol{R}^{-1}\boldsymbol{\alpha}\,\bar{\boldsymbol{\alpha}}_\eta^{\mathrm{H}}\boldsymbol{R}^{-1}\bar{\boldsymbol{\alpha}}_\eta\} \tag{6.60}$$

$$\boldsymbol{J}_{\eta\gamma} = \boldsymbol{J}_{\gamma\eta} = 2L\ (\sigma_s^2)^2\mathrm{Re}\{\boldsymbol{\alpha}^{\mathrm{H}}\boldsymbol{R}^{-1}\bar{\boldsymbol{\alpha}}_\eta\boldsymbol{\alpha}^{\mathrm{H}}\boldsymbol{R}^{-1}\bar{\boldsymbol{\alpha}}_\gamma + \boldsymbol{\alpha}^{\mathrm{H}}\boldsymbol{R}^{-1}\boldsymbol{\alpha}\,\bar{\boldsymbol{\alpha}}_\gamma^{\mathrm{H}}\boldsymbol{R}^{-1}\bar{\boldsymbol{\alpha}}_\eta\} \tag{6.61}$$

至此，FIM$\boldsymbol{J}$ 可用直接依赖于参数计算得到，目标的直达波和二维极化参数对应 CRB 为

$$\left.\begin{aligned}
\mathrm{CRB}(\theta_d) &= [\boldsymbol{J}^{-1}]_{1,1} \\
\mathrm{CRB}(\gamma) &= [\boldsymbol{J}^{-1}]_{2,2} \\
\mathrm{CRB}(\eta) &= [\boldsymbol{J}^{-1}]_{3,3}
\end{aligned}\right\} \tag{6.62}$$

§ 6.4　仿真结果分析

考虑该米波极化 MIMO 阵列雷达的阵元数 $M=10$，阵元间距为半波长布

局。入射频率为 300 MHz，入射波长 $\lambda=1$ m，目标直达波角度为 4°，反射角角度根据公式(5.13)计算得到，极化辅角和极化相位差设置为 $\gamma=45°$，$\eta=90°$，信号为左旋圆极化信号。信噪比 SNR=20 dB，500 个蒙特卡洛实验，快拍数 10 个。其中天线高度 $h_a=5$ m，目标高度 $h_t=7\ 000$ m，设置盐水场景，根据表 6.1，则可设置反射系数中的介电常数 $\varepsilon_r=75$ 和表面物质传导率 $\sigma_e=0.5$。图 6.3 和图6.4 给出了广义 MUSIC 和导向矢量合成降维 MUSIC 的 10 次谱估计结果。

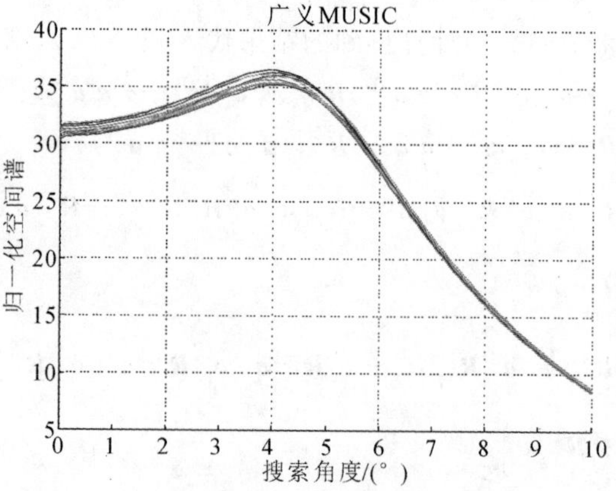

图 6.3　广义 MUSIC 的 10 次谱估计结果

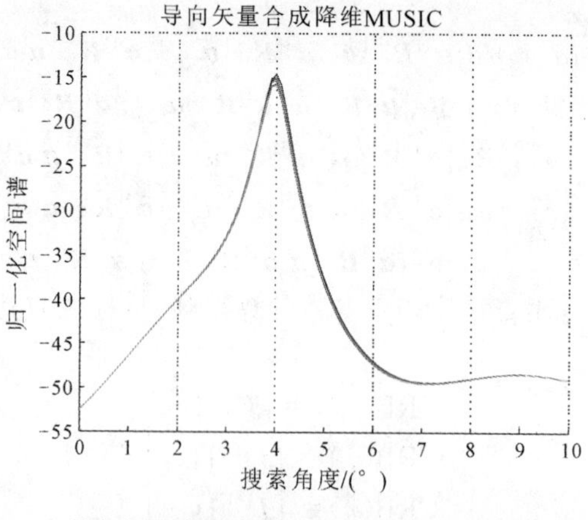

图 6.4　导向矢量合成降维 MUSIC 的 10 次谱估计结果

考虑该米波极化 MIMO 阵列雷达的阵元数 $M=10$，阵元间距为半波长布局。入射频率为 300 MHz，入射波长 $\lambda=1$ m，目标直达波角度为 4°，反射角角度根据公式(5.13)计算得到，极化辅角和极化相位差设置为 $\gamma=45°$，$\eta=90°$。信噪比变化，500 个蒙特卡洛实验，快拍数 10 个。其中天线高度 $h_a=5$ m，目标高度 $h_t=7\ 000$ m，设置盐水场景，根据表 6.1，则可设置反射系数中的介电常数 $\varepsilon_r=75$ 和表面物质传导率 $\sigma_e=0.5$。图 6.5 给出广义 MUSIC 和导向矢量合成降维 MUSIC 算法的直达波角度估计的 RMSE 和 CRB。从图中可以看出，在低信噪比下，广义 MUSIC 比导向矢量合成降维 MUSIC 算法估计结果要好，信噪比较高时，反之。此外，两种算法离最佳性能均有一定距离。图 6.6 给出导向矢量合成降维 MUSIC 算法对两种极化参数的估计结果，从图中可以看出，其能够较好地估计二维极化参数，且能够接近最优性能。

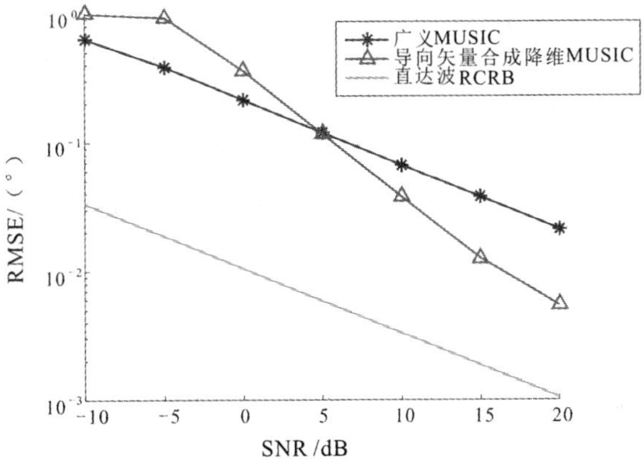

图 6.5　本章所提两种算法的 RMSE 和 CRB

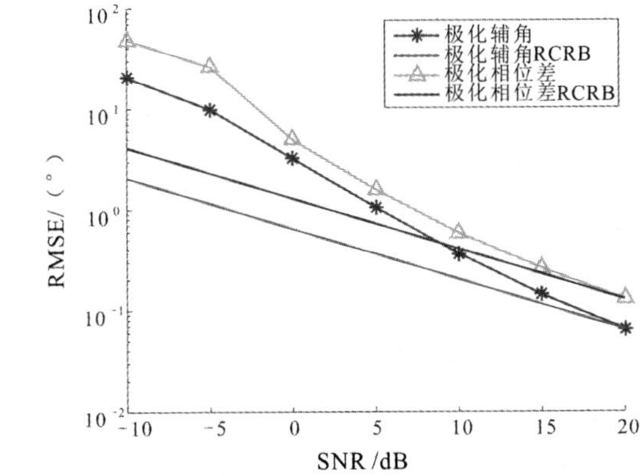

图 6.6　本章所提导向矢量合成降维 MUSIC 算法的极化参数估计 RMSE 和 CRB

考虑该米波极化 MIMO 阵列雷达的阵元数 $M=10$，阵元间距为半波长布局。入射频率为 300 MHz，入射波长 $\lambda=1$ m，其中天线高度 $h_a=5$ m，目标高度 $h_t=7\ 000$ m，设置盐水场景，根据表 6.1，则可设置反射系数中的介电常数 $\varepsilon_r=75$ 和表面物质传导率 $\sigma_e=0.5$。极化辅角和极化相位差分别设置为 $\gamma=45°$，$\eta=90°$。模拟设置目标从 50 km 飞出到 100 km 处的航迹，信噪比等于 10 dB，500 个蒙特卡洛实验，快拍数 10 个。图 6.7 给出广义 MUSIC 和导向矢量合成降维 MUSIC 算法的航迹跟踪图。图 6.7(a)为两种算法仰角测量结果与目标真实仰角的对比图。图 6.7(b)为两种算法仰角测量误差结果。图 6.7(c)为两种算法高度测量结果与目标真实高度的对比图。图 6.7(d)为两种算法高度测量误差结果。从图 6.7 中可看出，本章可对任意仰角的目标进行正确的估计，且两种算法随着目标的仰角变低，测量的误差逐渐变大，这与预期结果相吻合。

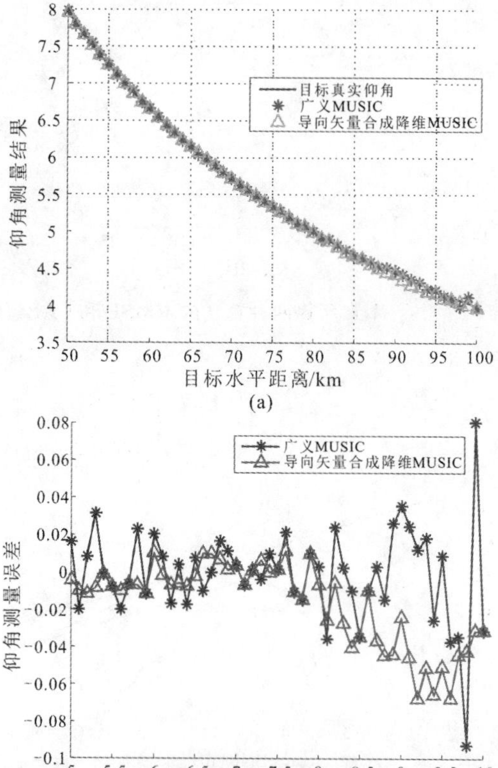

图 6.7　模拟航迹的跟踪测量结果

(a)目标的仰角测量结果；(b)目标的仰角测量误差结果；

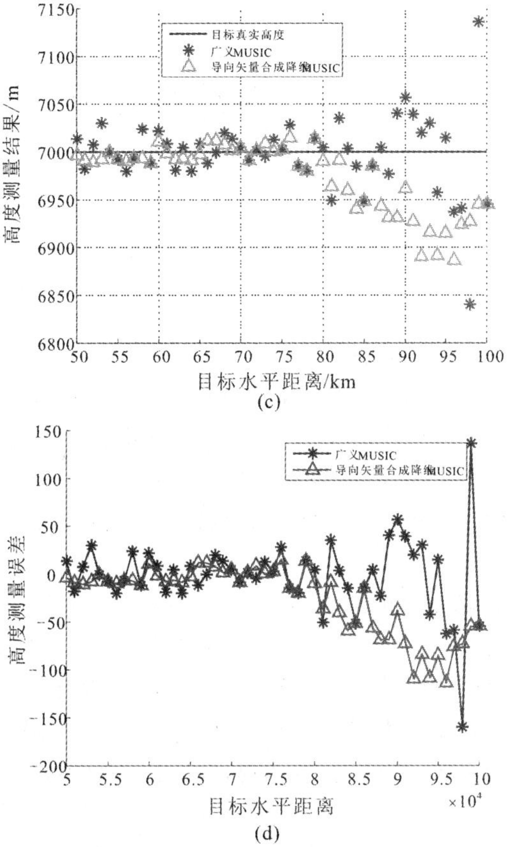

续图 6.7　模拟航迹的跟踪测量结果

(c)目标的高度测量结果;(d)目标的高度测量误差结果

§6.5　本 章 小 结

本章首先比较完善地推导了米波极化 MIMO 雷达的低仰角区域测高模型，并将接收到的数据进行适当的归类与变形，然后利用 MUSIC 算法得到低仰角区域目标的测高结果。其中，MUSIC 算法分为两类：一类是广义 MUSIC 算法，该算法并不涉及反射系数，对反射系数具有较好的鲁棒性，但是该算法以牺牲极化角度来达到目的；另一类算法是导向矢量合成降维 MUSIC 算法，该算法能够同时得到目标仰角和极化参数，但是需要提前知道反射系数。本算法测高结果距离最优性能还有一定的差距，极化信息能够接近最优 CRB。

第 7 章　总结及展望

§7.1　总　　结

目标参数估计是雷达和通信系统的重要任务之一。本书在常规阵列的基础上研究基于极化分集的极化敏感阵列和波形分集的 MIMO 阵列雷达的目标参数估计算法。本书的研究是原创性工作,并不属于开创性工作,针对原有研究成果中存在的部分问题提出的解决方案,可为基于分集技术的阵列雷达和通信系统提供理论支撑。主要解决的参数估计问题和其相对应的方案有四个部分:

第一个部分要解决的问题是,极化分集的极化敏感阵列若是由短电偶极子和小半径磁环构成,则阵列中的阵元辐射效率太低,这在雷达和通信系统中的某些应用是需要改进的部分。很显然,可用长电偶极子和大半径磁环组成的阵列来提高辐射效率。本书是在前人的工作基础之上提出一种新的长电偶极子和大半径磁环组成的极化敏感阵列,并提出一种不需要先验知识和基于 ESPRIT 的参数估计算法。该阵列可由三正交的长电偶极子组成,或者由三正交的大半径磁环组成,或者由混编的长电偶极子和大半径磁环组成。该阵列共有八个阵元,分为两个完全相同的子阵。很明显可利用其空域旋转不变性求解得到周期性模糊二维方向余弦的精估计,并通过相位补偿得到三正交共点式电偶极子矢量,然后得到二维角度模糊粗估计。将同时模糊的精、粗两组估计值做差值对比,找到差值最小的一组估计值,将该组的精估计值作为目标的最终二维角度精估计值,此时角度估计值无模糊且精度高。该算法无需搜索,计算量小。

第二部分解决了两个问题:一是 FDA-MIMO 雷达角度和距离联合估计时存在的距离模糊问题;二是速度矢量传感器 MIMO 阵列中的相干源目标 DOA 估计问题。针对第一个问题,本书提出利用参差频率增量来扩大距离模糊范围,并对二维参数估计问题采用 ESPRIT 和 MUSIC 联合一维搜索算法来降低计算复杂度。首先了解距离模糊的原因是 FDA 的均匀加权发射方向图是一个类 sinc 函数,因此它在距离维上是有栅瓣的。然后根据脉冲雷达信号处理中采用参差频率解模糊的思想，将其移植到本书的角度和距离联合参数估计中，即发射参差频率增量来提高出现栅瓣的距离。针对二维参数估计算法，先利用 ESPRIT 算法将角度估计出来，然后用得到的角度值回代入 MUSIC 算法进行一维距离搜索即可得到无模糊距离值，且角度和距离估计值是自动配对的。针对速度传感器 MIMO 阵列相干源二维 DOA 估计问题，提出一种新的 VFS 预处理方法去除信源的相干性。首先将所有速度向量中方向相同的传感器划分为同一个子阵列，则可得到三个空域完全相同的子阵列，然后利用矢量天线之间的速度场不同的信息进行平滑处理，以恢复 SCM 的秩，并计算了 VFS 预处理后 SCM 的互相关系数，以分析其去相关性能。本书提出的去相关算法有以下特点：不需要速度矢量传感器的位置信息，适用于任意构型阵列，不存在有效孔径的损失问题。

第三部分的内容是研究波形分集的米波 MIMO 雷达的低俯仰角参数估计问题，即传统米波雷达中的测高问题。米波雷达是一种反隐身体制雷达，但其孔径有限、多径等导致其测高精度不甚理想。MIMO 雷达匹配滤波后具有增强孔径的特点，能够提高参数估计的分辨力和估计精度。针对米波 MIMO 雷达测高问题，最大似然和广义 MUSIC 算法是米波 MIMO 阵列雷达测高方法行之有效的算法，但其计算量大，工程中难以接受。本书提出一种 BOMP 预处理的方法来降低搜索范围，从而降低上述两种有效参数估计算法的计算复杂度。首先对 MIMO 阵列接收数据稀疏化处理，通过数学操作将其变形至适合于 BOMP 算法的信号模型，然后利用粗栅格得到角度粗估计，并以此为初始值中心，取 MIMO 雷达波束宽度作为搜索范围。计算机仿真结果表明该算法能有效降低搜索类测高算法的计算复杂度。此外，针对复杂阵地，增强算法鲁棒性是一种行之有效的思路。但是笔者认为建立精细化测高模型才是解决问题的根本，为此建立起复杂阵地条件下的反射系数、反射高度、反射点位置三维联动的精确模型，然后将广义 MUSIC 算法和最大似然估计算法应用于精确地形信号模型下的目标高度测量。仿真结果表明精确信号模型的地形匹配估计结果比传统的鲁棒性地形失配估计算法精度要高。

第四部分研究了极化分集和波形分集相结合的米波极化 MIMO 雷达的测高问题。单独利用分集技术来测高已有研究成果，但是极化 MIMO 雷达的测高模型相对复杂，本书首先推导两种分集技术相结合的平坦镜面反射的测高模型。针对该测高信号模型，传统的基于 MUSIC 的算法并不可直接应用，故对本书信号模型进行适当的归类和变形，使得能够适用于经典超分辨 MUSIC 测高算法。通过进一步推导得到更加有效的广义 MUSIC 算法和导向矢量合成 MUSIC 算法。此外,给出两种算法的计算复杂度和极化 MIMO 雷达测高的 CRB 推导结果。最后利用仿真结果验证了算法的有效性。

§7.2 不足与展望

本书围绕极化分集的极化敏感阵列和波形分集的 MIMO 阵列雷达参数估计方面的问题进行了理论研究,解决了当前存在的一些问题,但仍存有一些重要的问题需要研究者去深入挖掘并解决。下面阐述本书存在的一些不足和未来研究的展望:

(1)对参数估计中存在阵元位置误差、极化 MIMO 互耦、通道的一致性等不理想的条件没有考虑,对参数估计的鲁棒性算法需要深入考虑,尤其是米波 MIMO 和米波极化 MIMO 雷达复杂地形的情况下,测高的精度急剧下降,鲁棒性算法显得非常重要。

(2)目前,已经存在不少极化敏感阵列系统和 MIMO 雷达系统,本书算法并没有利用实际系统进行验证,希望在未来可以进行实际系统的验证。

(3)米波极化 MIMO 雷达在复杂阵地下的测高问题将是接下来的工作重点,事实上针对平坦阵地,现有的算法已经能够较好地解决测高问题,但仍存在地形复杂,信号反射情况复杂,存在几个多径反射、反射规律依然不是很清楚的问题,因此算法的鲁棒性是需要深入研究的课题。

附录 本书符号和缩略语说明

一、本书符号

本书中斜体加粗的小写字母表示向量，斜体加粗的大写字母表示矩阵，j 表示虚数，π 表示圆周率。

书中主要运算符号的含义见附表 1。

附表 1 本书符号的含义

符 号	说 明
$(\cdot)^{T}$	矩阵或者向量的转置
$(\cdot)^{H}$	矩阵或者向量的复共轭转置
$(\cdot)^{*}$	复共轭运算
$\odot$	点积，两个向量对应元素相乘
$\otimes$	Kronecker 内积
$\oplus$	Hadamard 乘积
$\mathrm{Tr}(\cdot)$	求矩阵迹，矩阵对角线元素之和
$\mathrm{diag}[\cdot]$	矩阵对角化，矢量元素构成对角矩阵
$\mathrm{vec}(\cdot)$	矩阵矢量化，按列将矩阵元素依次排列
$\min(\cdot)$	求极小化操作
$\max(\cdot)$	求极大化操作
$E\{\cdot\}$	求数学期望
$\angle(\cdot)$	取相位操作，范围是 $-\pi \sim \pi$
$[\cdot]_{m}$	取方括号内矢量的第 m 元素
$[\cdot]_{m,m}$	取方括号内矢量的第 m 元素

续 表

符　号	说　明
$\lfloor \cdot \rfloor$	向下取整
$\lceil \cdot \rceil$	向上取整
$\boldsymbol{I}_m$	维数为 m 的单位矩阵
$\mathbf{0}_{m\times n}$	维数为 $m\times n$ 的全零矩阵
$e^{(\cdot)}$ 或 $\exp(\cdot)$	以自然数为底的指数
$\mathrm{rank}[\cdot]$	矩阵的秩
$J_1(\cdot)$	第一类贝塞尔函数，且阶数为1
$\det(\cdot)$	计算矩阵行列式

二、缩略语

本书正体字母组合表示缩略语，英文缩写、英文全称与中文全称见附表2。

附表2　缩略语

缩　写	英　文	中　文
AS-SCM	After-Smoothing SCM	平滑后协方差矩阵
BOMP	Block Orthogonal Matching Pursuit	块正交匹配追踪
CRB	Cramer-Rao Bound	克拉美-罗界
DOA	Direction-of-Arrival	波达方向(接收角)
DOD	Direction-of-Departure	波离方向(发射角)
ESPRIT	Estimation of Signal Parameter via Rotational Invariance Technique	旋转不变子空间算法
FIM	Fisher Information Matrix	费舍尔信息矩阵
FDA	Frequency Diversity Array	频率分集阵列
MIMO	Multiple-Input Multiple-Output	多输入多输出
MTD	Moving Target Detection	动目标检测

续　表

缩　写	英　文	中　文
MTI	Moving Target Indication	动目标显示
MUSIC	Multiple Signal Classification	多重信号分类算法
OMP	Orthogonal Matching Pursuit	正交匹配追踪
RCRB	Root-CRB	求根克拉美-罗界
RMSE	Root Mean Squared Error	均方根误差
RCS	Radar Cross Section	雷达散射截面积
SNR	Signal to Noise Ratio	信噪比
SCM	Source Covariance Matrix	信源协方差矩阵
VFS	Velocity-Field Smoothing	速度场平滑

参考文献

[1] FERRARA E, PARKS T. Direction finding with an array of antennas having diverse polarizations [J]. IEEE Transactions on Antennas and Propagation, 1983, 31(2): 231 - 236.

[2] LI J, COMPTON R T. Angle and polarization estimation using ESPRIT with a polarization sensitive array [J]. IEEE Transactions on Antennas and Propagation, 1991, 39(9): 1376 - 1383.

[3] LI J, COMPTON R T. Two-dimensional angle and polarization estimation using the ESPRIT algorithm [J]. IEEE Transactions on Antennas and Propagation, 1992, 40(5): 550 - 555.

[4] LI J, COMPTON R T. Angle estimation using a polarization sensitive array [J]. IEEE Transactions on Antennas and Propagation, 1991, 39 (10): 1539 - 1543.

[5] LI J, COMPTON R T. Angle and polarization estimation in a coherent signal environment [J]. IEEE Transactions on Aerospace and Electronic Systems, 1993, 29(3): 706 - 716.

[6] WEISS A J, FRIEDLANDER B. Performance analysis of diversely polarized antenna arrays [J]. IEEE Transactions on Signal Processing 1991, 39(7): 1589 - 1603.

[7] FRIEDLANDER B, WEISS A J. Performance of diversely polarized antenna arrays for correlated signals [J]. IEEE Transactions on Aerospace and Electronic Systems, 1992, 38(3): 869 - 879.

[8] WEISS A J, FRISDLANDER B. A direction finding algorithm for diversely polarized arrays [J]. Digital Signal Processing, 1992, 2(3): 123 - 134.

[9] WEISS A J, FRIEDLANDER B. Direction finding for diversely polarized signals using polynomial rooting [J]. IEEE Transactions on Signal Processing, 1993, 41(5): 1893 - 1905.

[10] WEISS A J, FRIEDIANDER B. Analysis of a signal estimation algorithm for diversely polarized arrays [J]. IEEE Transactions on Signal Processing,

1993,41(8):2628－2638.

[11] WEISS A J, FRIEDIANDER B. Maximum likelihood signal estimation for polarization sensitive arrays [J]. IEEE Transactions on Signal Processing, 1993,41(7):918－925.

[12] NEHORAI A, PALDI E. Vector-sensor array processing for electromagnetic source localization [J]. IEEE Transactions on Signal Processing, 1994,42(2):376－398.

[13] HO K C, TAN K C, SER W. An investigation on number of signals whose directions of arrival are uniquely determinable with an electromagnetic vector sensor [J]. Signal Processing, 1995, 47 (1):41－54.

[14] HOCHWALD B, NEHORAI A. Identifiability in array processing models with vector-sensor applications [J]. IEEE Transactions on Signal Processing, 1996,44(1):83－95.

[15] TAN K C, HO K C, NEHORAI A. Linear independence of steering vectors of an electromagnetic vector sensor [J]. IEEE Transactions on Signal Processing, 1996,44(12):3099－3107.

[16] TAN K C, HO K C, NEHORAI A. Uniqueness study of measurements obtainable with arrays of electromagnetic vector sensors [J]. IEEE Transactions on Signal Processing, 1996,44(4):1036－1039.

[17] HO K C, TAN K C, TAN B T G. Efficient method for estimating directions-of-arrival of partially polarized signals with electromagnetic vector sensors [J]. IEEE Transactions on Signal Processing, 1997,45 (10):2485－2497.

[18] NEHORAI A, TICHAVSKY P. Cross-product algorithms for source tracking using an EM vector sensor [J]. IEEE Transactions on Signal Processing, 1999,47(10):2863－2867.

[19] KO C C, ZHANG J, NEHORAI A. Separation and tracking of multiple broadband sources with one electromagnetic vector sensor [J]. IEEE Transactions on Aerospace and Electronic Systems, 2002,38(3): 1109－1116.

[20] SEE C M S, NEHORAI A. Source localization with distributed electromagnetic component sensor array processing [C]//IEEE

Seventh International Symposium on Signal Processing and Its Applications, IEEE Press, 2003, 1: 177 - 180.

[21] HURTADO M, NEHORAI A. Performance analysis of passive low-grazing-angle source localization in maritime environments using vector sensors [J]. IEEE Transactions on Aerospace and Electronic Systems, 2007, 43(2): 780 - 789.

[22] MONTE L L, ELNOUR B, ERRICOLO D, et al. Design and realization of a distributed vector sensor for polarization diversity applications [C]//IEEE International Waveform Diversity and Design Conference. Pisa: IEEE Press, 2007: 358 - 361.

[23] MONTE L L, EINOUR B, ERRICOLO D. Distributed 6D vector antennas design for direction of arrival application [C]//IEEE International Conference on Electromagnetics in Advanced Applications. Torino: IEEE Press, 2007: 431 - 434.

[24] MONTE L L, ELNOUR B, RAJAGOPALAN A, et al. Circularly and linearly distributed narrowband vector antennas for direction of arrival applications [C]//North American Radio Science Conference. Ottawa, Ontario, Canada, 2007: 22 - 26.

[25] TABRIKIAN J, SHAVI R, RAHAMIM D. An efficient vector sensor configuration for source localization [J]. IEEE singal Processing Letters, 2004, 11(8): 690 - 693.

[26] RAHAMIM D, TABRIKIAN J, SHAVIT R. Source localization using vector sensor array in a multipath environment [J]. IEEE Transactions on Signal Processing, 2004, 52(11): 3096 - 3103.

[27] BIHAN N L, MARS J. Singular value decomposition of quaternion matrices: a new tool for vector-sensor signal processing [J]. Signal Processing, 2004, 84(7): 1177 - 1199.

[28] MIRON S, BIHAN N L, MARS J. Quaternion-MUSIC for vector-sensor array processing [J]. IEEE Transactions on Signal Processing, 2006, 54(4): 1218 - 1229.

[29] BIHAN N L, MIRON S, MARS J. MUSIC algorithm for vector-sensors array using biquaternions [J]. IEEE Transactions on Signal Processing, 2007, 55(9): 4523 - 4533.

[30] WONG K T, ZOLTOWSKI M D. Uni-vector-sensor ESPRIT for multi-source azimuth, elevation, and polarization estimation [J]. IEEE Transactions on Antennas and Propagation, 1997, 45(10): 1467 - 1474.

[31] WONG K T, ZOLTOWSKI M D. Self-initiating MUSIC direction finding and polarization estimation in spatio-polarizational beamspace [J]. IEEE Transactions on Antennas and Propagation, 2000, 48(8): 1235 - 1245.

[32] ZOLTOWSKI M D, WONG K T. ESPRIT-based 2D direction finding with a sparse array of electromagnetic vector-sensors [J]. IEEE Transactions on Signal Processing, 2000, 48(8): 2195 - 2204.

[33] ZOLTOWSKI M D, WONG K T. Closed-form eigenstructure-based direction finding using arbitrary but identical subarrays on a sparse uniform rectangular array grid [J]. IEEE Transactions on Signal Processing, 2000, 48(8): 2205 - 2210.

[34] WONG K T, ZOLTOWSKI M D. Closed: form direction-finding with arbitrarily spaced electromagnetic vector-sensors at unknown locations [J]. IEEE Transactions on Antennas and Propagation, 2000, 48(5): 671 - 681.

[35] WONG K T. Direction finding/polarization estimation: dipole and/or loop triad(s) [J]. IEEE Transactions on Aerospace and Electronic Systems, 2001, 37(2): 679 - 684.

[36] WONG K T, LI L, ZOLTOWSKI M D. Root-MUSIC-based direction-finding and polarization estimation using diversely polarized possibly collocated antennas [J]. IEEE Antennas and Wireless Propagation Letters, 2004, 3: 129 - 132.

[37] LUO F, YUAN X. Enhanced "vector-cross-product" direction-finding using a constrained sparse triangular-array [J]. EURASIP Journal on Advances in Signal Processing, 2012, 2012: 115.

[38] YUAN X. Estimating the DOA and the polarization of a polynomial-phase signal using a single polarized vector-sensor [J]. IEEE Transactions on Signal Processing, 2012, 60(3): 1270 - 1282.

[39] YUAN X, WONG K T, AGRAWAL K. Polarization estimation with a dipole-dipole pair, a dipole-oop pair, or a loop-loop pair of various

orientations [J]. IEEE Transactions on Antennas and Propagation, 2012,60(5):2442-2452.

[40] YUAN X, WONG K T, XU Z, et al. Various triads of collocated dipoles/loops, for direction finding & polarization estimation [J]. IEEE Sensors Journal, 2012,12(6):1763-1771.

[41] YUAN X. Quad compositions of collocated dipoles and loops: for direction finding and polarization estimation [J]. IEEE Antennas and Wireless Propagation Letters, 2012,11:1044-1047.

[42] YUAN X. Diversely polarized antenna-array signal processing [D]. [Ph. D. dissertation], Hong Kong: The Hong Kong Polytechnic University, 2012.

[43] WONG K T, YUAN X. "Vector cross-product direction-inding" with an electromagnetic vector-sensor of six orthogonally oriented but spatially noncollocating dipoles/loops [J]. IEEE Transactions on Signal Processing, 2011,59(1):160-171.

[44] SONG Y, WONG K T, YUAN X. Correction to Vector cross-product direction-finding" with an electromagnetic vector-sensor of six orthogonally oriented but spatially noncollocating dipoles/loops [J]. IEEE Transactions on Signal Processing, 2014,62(4):1028-1030.

[45] WONG K T, SONG Y, FULTON C J, et al. Electrically "Long" Dipoles in a collocated/orthogonal triad: for direction finding and polarization estimation [J]. IEEE Transactions on Antennas and Propagation, 2017,65(11):6057-6067.

[46] KHAN S, WONG K T, SONG Y, et al. Electrically large circular loops in the estimation of an incident emitter's direction-of-arrival or polarization [J]. IEEE Transactions on Antennas and Propagation, 2018,66(6):3046-3055.

[47] KHAN S, WONG K T. A six - component vector sensor comprising electrically long dipoles and large loops: to simultaneously estimate incident sources' direction - of - arrival and polarizations [J]. IEEE Transactions on Antennas and Propagation, 2020,68(8):6355-6363.

[48] 徐振海. 极化敏感阵列信号处理的研究[D]. 长沙:国防科学技术大学,2004.

[49] 庄钊文,徐振海,肖顺平,等.极化敏感阵列信号处理[M].北京:国防工业出版社,2005.

[50] 徐振海,王雪松,冯德军,等.极化域-空域动态联合谱估计[J].电波科学学报,2005,20(1):25－28.

[51] 徐振海,王雪松,肖顺平,等.极化域-空域联合谱估计[J].国防科技大学学报,2004,26(3):63－67.

[52] 徐振海,肖顺平,王雪松,等.极化域-空域联合谱估计精度研究[J].信号处理,2006,22(3):317－320.

[53] 徐振海,肖顺平,王雪松,等.极化域-空域联合谱分辨力研究[J].信号处理,2008,24(1):7－10.

[54] 徐友根,刘志文,龚晓峰.极化敏感阵列信号处理[M].北京:北京理工大学出版社,2013.

[55] 徐友根,刘志文,王四平.二维正交矢量天线导向矢量的秩-1模糊问题研究[J].电路与系统学报,2006,11(2):20－23.

[56] 徐友根,刘志文,王四平.三维正交矢量天线导向矢量的秩-1模糊[J].系统工程与电子技术,2005,27(1):1－5.

[57] 徐友根,刘志文,王四平.四维正交矢量天线导向矢量的秩-1模糊[J].信号处理,2005,21(4A):48－52.

[58] 徐友根,刘志文,王四平.五维正交矢量天线导向矢量的秩-1模糊问题研究[J].电子与信息学报,2005,27(5):749－752.

[59] 徐友根,刘志文.电磁矢量传感器及其阵列累量域虚拟导向矢量的线性无关度[J].电子与信息学报,2005,27(6):983－986.

[60] XU Y,LIU Z,WONG K T,et al.Virtual-manifold ambiguity in HOS-based direction-finding with electromagnetic vector-sensors [J].IEEE Transactions on Aerospace and Electronic Systems,2008,44(4):1291-1308.

[61] 徐友根,刘志文. 电磁矢量传感器阵列相干信号源DOA和极化参数的同时估计:空间平滑方法[J].通信学报,2004,25(5):28－38.

[62] XU Y, LIU Z. Polarimetric angular smoothing algorithm for an electromagnetic vector-sensor array [J].IET Radar Sonar Navigation,2007,1(3):230－240.

[63] 徐友根,刘志文.基于累积量的极化敏感阵列信号DOA和极化参数的同时估计[J].电子学报,2004,32(12):1962－1966.

[64] 徐友根,刘志文.广义信号子空间拟合角度-极化联合估计[J].北京理工大学学报,2010,30(7):835-839.

[65] 龚晓峰,刘志文,徐友根.电磁矢量传感器阵列信号DOA估计:双模MUSIC[J].电子学报,2008,36(9):1698-1673.

[66] 龚晓峰,徐友根,刘志文.四元数域低秩逼近及其在矢量阵列DOA估计中的应用[J].北京理工大学学报,2008,28(11):1013-1017.

[67] GONG X F, LIU Z W, XU Y G. Quad-quaternion MUSIC for DOA Estimation using electromagnetic vector sensors [J]. EURASIP Journal on Advances in Signal Processing, 2008, 2008(23): 1-14.

[68] GONG X F, LIU Z W, XU Y G. Regularised parallel factor analysis for the estimation ofdirection-of-arrival and polarisation with a single electromagnetic vector-sensor [J]. IET Signal Processing, 2011, 5(4): 390-396.

[69] GONG X F, LIU Z W, XU Y G. Direction finding via biquaternion matrix diagonalization with vector-sensors [J]. Signal Processing, 2011, 91(4): 821-831.

[70] GONG X F, WANG K, LIN Q H. Simultaneous source localization and polarization estimation via non-orthogonal joint diagonalization with vector-sensors [J]. Sensors, 2012, 12: 3394-3417.

[71] GONG X F, LIU Z W, XU Y G. Coherent source localization: bicomplex polarimetric smoothing with electromagnetic vector-sensors [J]. IEEE Transactions on Aerospace and Electronic Systems, 2011, 47(3): 2268-2285.

[72] GONG X F, JIANG J C, LI H, et al. Spatially spread dipole/loop quint for vector-cross-product-based direction finding and polarisation estimation [J]. IET Signal Processing, 2018, 12(5): 636-642.

[73] HE J, LIU Z. Computationally efficient 2D direction finding and polarization estimation with arbitrarily spaced electromagnetic vector sensors at unknown locations using the propagator method [J]. Digital Signal Processing, 2009, 19(3): 491-503.

[74] HE J, JIANG S, WANG J, et al. Polarization difference smoothing for direction finding of coherent signals [J]. IEEE Transactions on Aerospace and Electronic Systems, 2010, 46(1): 469-480.

[75] GU C, HE J, ZHU X, et al. Efficient 2D DOA estimation of coherent signals in spatially correlated noise using electromagnetic vector sensors [J]. Multidimensional Systems and Signal Processing, 2010, 21:239 - 254.

[76] LIU Z, HE J, LIU Z. Computationally efficient DOA and polarization estimation of coherent sources with linear electromagnetic vector-sensor array [J]. EURASIP Journal on Advances in Signal Processing, 2011(1):1 - 10.

[77] JIANG J F, ZHANG J Q. Geometric algebra of Euclidean 3-space for electromagnetic vector-sensor array processing: Part I—Modeling [J]. IEEE Transactionson Antennas and Propagation, 2010, 58(12): 3961 - 3973.

[78] JIANG J F, ZHANG J Q. A weighted inner product estimator in the geometric algebra of euclidean 3-space for source localization using an EM vector-sensor [J]. Chinese Journal of Aeronautics, 2012, 25(1): 83 - 93.

[79] Li Y, ZHANG J Q. An Enumerative NonLinear Programming approach to direction finding with a general spatially spread electromagnetic vector sensor array [J]. Signal Processing, 2013, 93(4):856 - 865.

[80] 王兰美，黄际英，廖桂生.部分极化波参数估计方法的改进[J].电子与信息学报，2006，28(9)：1603 - 1606.

[81] 王洪洋，王兰美，廖桂生.基于单矢量传感器的信号多参数估计方法[J].电波科学学报，2005，20(1)：15 - 19.

[82] 王洪洋，廖桂生，王兰美.基于矢量传感器阵列的信号多参数估计方法[J].电路与系统学报，2006，11(3)：88 - 95.

[83] 王兰美.极化阵列的参数估计和滤波方法研究[D].西安：西安电子科技大学，2005.

[84] YANG M L, DING J, CHEN B, et al. A multiscale sparse array of spatially spread electromagnetic-vector-sensors for direction finding and polarization estimation [J]. IEEE Access, 2018, 6(99):9807 - 9818.

[85] 张小飞，汪飞，陈伟华，等.阵列信号处理的理论与应用[M].北京：国防工业出版社，2013.

[86] 张小飞,陈华伟,仇小锋,等.阵列信号处理及MATLAB实现[M].北京:电子工业出版社,2015.

[87] ZHANG X F, XU D Z. Novel blind joint direction of arrival and polarization estimation for polarization sensitive uniform circular array [J].Progress in Electromagnetics Research,2008,86:19-37.

[88] ZHANG X F, CHEN C, LI J, et al. Blind DOA and polarization estimation for polarization-sensitive array using dimension reduction MUSIC[J]. Multidimensional Systems and Signal Processing, 2014, 25(1):67-82.

[89] 黄家才,石要武,陶建武,等.宽带循环平稳信号二维DOA和极化参数联合估计算法[J].系统仿真学报,2007,19(5):1100-1103.

[90] 黄家才,石要武,陶建武.基于双电磁矢量传感器的近场源多参数估计[J].电波科学学报,2007,22(5):848-854.

[91] 李京书,陶建武.信号DOA和极化信息联合估计的降维四元数MUSIC方法[J].电子与信息学报,2011,33(1):106-111.

[92] TAO J W, LIU L, LIN Z Q. Joint DOA, range and polarization estimation of sources in the Fresnel region via a sparse line dual-polarization sensor array [J]. IEEE Transactions on Aerospace and Electronic Systems,2011,47(4):2657-2672.

[93] 王建英,陈天麒.频率、二维到达角和极化联合估计[J].电子学报,1999,27(11):74-76.

[94] 王建英,陈天麒.用四阶累积量实现频率、二维到达角和极化的联合估计[J].中国科学:E辑,2000,30(5):424-429.

[95] 王建英,王激扬,陈天麒.宽频段空间信号频率、二维到达角和极化联合估计[J].中国科学:E辑,2001,31(6):526-532.

[96] 司伟建,朱曈,张梦莹.平面极化天线阵列的DOA及极化参数降维估计方法[J].通信学报,2014,35(12):28-35.

[97] 吴娜,司伟建,焦淑红,等.利用极化敏感阵列特性的信源数估计技术研究[J].中南大学学报(自然科学版),2016(1):130-135.

[98] 司伟建,周炯赛,曲志昱.稀疏极化敏感阵列的DOA和极化参数联合估计[J].电子与信息学报,2016,38(5):1129-1134.

[99] 曾富红,曲志昱,司伟建.极化敏感阵列的DOA及极化参数降维估计算法[J].应用科技,2017,44(3):39-42.

[100] 黄小梅，司伟建，王虹，等.基于双旋转单偶极子阵列的谱参数估计方法[J].电波科学学报，2018(1)：77－84.

[101] 郑桂妹.电磁矢量传感器阵列的角度估计及其在MIMO雷达中的应[D].西安：西安电子科技大学，2014.

[102] 郑桂妹，陈伯孝，杨明磊，等.稀疏均匀非同心电磁矢量阵列的2D-DOA与极化联合估计[J].中国科学：F辑：信息科学，2014，44(9)：1171－1192.

[103] 郑桂妹，陈伯孝，吴渤.三正交分离式极化敏感阵列的DOA估计[J].电子与信息学报，2014，36(5)：1088－1093.

[104] 郑桂妹，陈伯孝，杨明磊.基于改进分离式电磁矢量传感器阵列的二维DOA估计方法 [J].电波科学学报，2014，29(2)：213－220.

[105] 郑桂妹，陈伯孝，杨明磊.新型拉伸电磁矢量传感器的二维高精度DOA估计[J].系统工程与电子技术，2014，36(7)：1282－1290.

[106] 郑桂妹，肖宇，宫健.基于分离式电磁矢量传感器阵列的相干信号DOA估计[J].系统工程与电子技术，2016，38(4)：753－759.

[107] ZHENG G M，WU B，MA Y，et al.DOA Estimation with a sparse uniform array of orthogonally oriented and spatially separated dipole-triads [J].IET Radar Sonar Navigation，2014，8(8)：885－894.

[108] ZHENG G M. Two-dimensional DOA estimation for polarization sensitive array consisted of spatially spread crossed-dipole [J]. IEEE Sensors Journal，2018，18(12)：5014－5023.

[109] LI B B，BAI W X，ZHENG G M.Successive ESPRIT algorithm for joint DOA-range-polarization estimation with polarization sensitive FDA-MIMO radar [J].IEEE Access，2018，6(1)：36376－36382.

[110] LI B B，BAI W X，ZHENG G M. High accuracy and unambiguous 2D-DOA estimation with an uniform planar array of“long”electric-dipoles [J].IEEE Access，2018，6(1)：40559－40568.

[111] LI B B，BAI W X，ZHENG G M.RAM-based angle estimation with linear spatially separated polarisation sensitive array [J]. International Journal of Electronics，2018，105(10)：1657－1672.

[112] LIU B，LI B B，FENG Y Q，et al. Vector cross product based 2D-DOA and polarization estimation with “long” electric-dipole quint [J].IEEE Access，2019，7：27075－27085.

[113] SONG Y W, HU G P, ZHENG G M, et al. ESPRIT-Based DOA Estimation with Spatially Spread Long Dipoles and/or Large Loops [J]. Circuits Systems And Signal Processing, 2020, 39(11): 5568 - 5587.

[114] 郑桂妹，张栋，张秦，等.基于分集技术的阵列多参数估计[M].北京：国防工业出版社，2020.

[115] FORSYTHE K W, BLISS D W, Fawcett G S. Multiple-Input Multiple-Output (MIMO) radar: performance issues [C]// Proceedings of IEEE Radar Conference, IEEE, 2004, 1: 310 - 315.

[116] ROBEY F C, COUTTS, WEIKLE D, et al. MIMO radar theory and experimental results [C]//Proceedings of the 38th Asilomar Conference on Signals, Systems and Computers, 2004, 1: 300 - 304.

[117] FISHLER E, HAIMOVICH A, BLUM R, et al. MIMO radar: an idea whose time has come[C]//In: Proc IEEE Radar Conf, Honolulu, Hawaii, USA, Apr. 2004, 2: 71 - 78.

[118] FORSYTHE K W, BLISS D W. Waveform correlation and optimization issues for MIMO radar [C]//Proceedings of the 39th Asilomar Conference on Signals, Systems and Computers, 2005: 1306 - 1310.

[119] FISHLER E, HAIMOVICH A, BLUM R L, et al. Spatial diversity in radars: models and detection performance[J]. IEEE Transactions on Signal Processing, 2006, 54(3): 823 - 838.

[120] LI J, STOICA P. MIMO radar with colocated antennas[J]. IEEE Signal Processing Magazine, 2007, 24(5): 106 - 114.

[121] LI J, STOICA P, XU L, et al. On parameter identifiability of MIMO radar[J]. IEEE Signal Processing Letters, 2007, 14(12): 968 - 971.

[122] HAIMOVICH A, BLUN R, CIMINI L. MIMO radar with widely separated antennas[J]. IEEE Signal Processing Magazine, 2008, 25(1): 116 - 129.

[123] XU L, LI J, STOICA P. Target detection and parameter estimation for MIMO radar systems[J]. IEEE Transactions on Aerospace and Electronic Systems, 2008, 44(3): 927 - 939.

[124] ZHANG Y, AMIN M G, HIMED B. Joint DOD/DOA estimation in MIMO radar exploiting time-frequency signal representations[J].

EURASIP Journal on Advances in Signal Processing,2012(1).DOI:10.1186/1687-6180-2012-102.

[125] ABOULNASR H, SERGIY A V. Transmit Energy Focusing for DOA Estimation in MIMO Radar With Colocated Antennas [J]. IEEE Transactions on Signal Processing,2011,59(6):2669-2682.

[126] ABOULNASR H, SERGIY A V, ARASH K. Transmit radiation pattern invariance in MIMO radar with application to DOA estimation[J].IEEE Signal Processing Letters,2015,22(10):1609-1613.

[127] ARASH K, ABOULNASR H, SERGIY A V, et al. Efficient Transmit beamspace design for search-free based DOA estimation in MIMO radar [J].IEEE Transactions on Signal Processing,2014,62(6):1490-1500.

[128] BEKKERMAN I, TABRIKIAN J. Target detection and localization using MIMO radars and sonars[J]. IEEE Transactions on Signal Processing,2006,54(10):3873-3883.

[129] BENCHEIKH M L, WANG Y, He H. Polynomial root finding technique for joint DOA DOD estimation in bistatic MIMO radar [J].Signal Processing,2010,90(9),2723-2730.

[130] BENCHEIKH M L, WANG Y. Joint DOD-DOA estimation using combined ESPRIT-MUSIC approach in MIMO radar [J].Electronics Letters,2010,46(15):1081-1083.

[131] CAO M Y, SERGIY A V, ABOULNASR H. Transmit array interpolation for DOA estimation via tensor decomposition in 2-D MIMO radar[J].IEEE Transactions on Signal Processing,2017,65(19):5225-5239.

[132] LIN Y C, LEE T S. Max-MUSIC: A low-complexity high-resolution direction finding method for sparse MIMO radars [J]. IEEE Sensors Journal,2020,20(24):14914-14923.

[133] SHI J P, HU G Q, ZHANG F, et al. Sparsity-based DOA estimation of coherent and uncorrelated targets with flexible MIMO radar[J]. IEEE Transactions on Vehicular Technology,2019,68(6):5835-5848.

[134] SHI J P, WEN F Q, LIU T P. Nested MIMO radar: coarrays, tensor modeling and angle estimation[J].IEEE Transactions on Aerospace

and Electronic Systems,2021:57(1):573－585.

[135] 郑志东,张剑云,杨瑛.基于发射波束域-平行因子分析的 MIMO 雷达收发角度估计 [J].电子与信息学报,2011,33(12):2875－2880.

[136] XU F,SERGIY V,YANG X P.Joint DOD and DOA Estimation in Slow-Time MIMO Radar via PARAFAC Decomposition[J]. IEEE Signal Processing Letters,2020,27,1495－1499.

[137] ZHANG W,LIU W,WANG J,et al.Joint Transmission and Reception Diversity Smoothing for DIRECTION Finding of Coherent Targets in MIMO Radar[J].IEEE Journal of Selected Topics in Signal Processing, 2014,8(1):115－124.

[138] 刘波.MIMO 雷达正交波形设计及信号处理研究[D].成都:电子科技大学,2007.

[139] XU J,WANG W Q,GUI R H.Computational efficient DOA,DOD, and doppler estimation algorithm for MIMO radar[J].IEEE Signal Processing Letters,2019,26(1):44－48.

[140] CHEN D F,CHEN B X,QIN G D.Angle estimation using ESPRIT in MIMO radar[J].Electronics Letters,2008,44(12):770－771.

[141] JUN M,LIAO G,Li J.Joint DOD and DOA estimation for bistatic MIMO radar [J].Signal Processing,2009,89(2):244－251.

[142] LIU N,ZHANG L R,ZHANG J.Direction finding of MIMO radar through ESPRIT and Kalman filter [J].Electronics Letters,2009,45 (17):908－910.

[143] ZHANG J, ZHANG L R, YANG Z W, et al. Signal subspace reconstruction method of MIMO radar [J]. Electronics Letters, 2010,46(7):531－533.

[144] YANG M L,CHEN B X,YANG X Y.Conjugate ESPRIT algorithm for bistatic MIMO radar [J].Electronics Letters,2010,46(25):1692－1694.

[145] XIE R,LIU Z,ZHANG Z.DOA estimation for monostatic MIMO radar using polynomial rooting [J].Signal Processing,2010,90(12): 3284－3288.

[146] LIU J,LIU Z,XIE R.Low angle estimation in MIMO radar [J]. Electronics Letters,2010,46(23):1565－1566.

[147] 谢荣.MIMO 雷达角度估计算法研究 [D].西安:西安电子科技大

学,2011.
[148] 刘晓莉,廖桂生.多基线数据融合的双基地 MIMO 雷达角度估计[J].电波科学学报,2010,25(6):1199-1205.
[149] 刘晓莉,廖桂生.基于 MUSIC 和 ESPRIT 的双基地 MIMO 雷达角度估计算法[J].电子与信息学报,2010,32(9):2179-2182.
[150] 谢荣,刘峥,刘韵佛.基于 L 型阵列 MIMO 雷达的多目标分辨和定位[J].系统工程与电子技术,2010,32(1):49-52.
[151] 谢荣,刘峥.基于多项式求根的双基地 MIMO 雷达多目标定位方法[J].电子与信息学报,2010,32(9):2197-2220.
[152] LI C,LIAO G,ZHU S,et al.An ESPRIT-like algorithm for coherent DOA estimation based on data matrix decomposition in MIMO radar[J].Signal Processing,2011,91(8):1803-1811.
[153] XIE R,LIU Z,WU J X.Direction finding with automatic pairing for bistatic MIMO radar [J].Signal Processing,2012,92(1):198-203.
[154] 符渭波,苏涛,赵永波,等.空间色噪声环境下基于时空结构的双基地 MIMO 雷达角度和多普勒频率联合估计方法[J].电子与信息学报,2011,33(7):1649-1654.
[155] 符渭波,苏涛,赵永波,等.空间色噪声环境下双基地 MIMO 雷达角度和多普勒频率联合估计方法 [J].电子与信息学报,2011,33(12):2858-2862.
[156] 符渭波,苏涛,赵永波,等.双基地 MIMO 雷达相干源角度估计方法[J].西安电子科技大学学报,2012,39(2):143-152.
[157] 张娟,张林让,刘楠,等.一种有效的 MIMO 雷达相干信源 DOA 估计方法[J].电子学报,2011,39(3):680-684.
[158] LIU Y,JIU B,XIA G,et al.Height measurement of low-angle target using MIMO radar under multipath interference[J].IEEE Transactions on Aerospace and Electronic Systems,2018,54(2):808-818.
[159] YANG M,SUN L,YUAN X,et al.A new nested MIMO array with increased degrees of freedom and hole-free difference coarray[J].IEEE Signal Processing Letters,2018,25(1):40-44.
[160] ZHANG X,XU D.Angle estimation in MIMO radar using reduced-dimension Capon [J].Electronics Letters,2010,46(12):860-861.
[161] ZHANG X,XU D.Low-complexity ESPRIT-based DOA estimation

for colocated MIMO radar using reduced-dimension transformation [J].Electronics Letters,2011,47(4):283 - 284.

[162] ZHANG X,XU L,XU L,et al.DOD and DOA estimation in MIMO radar with reduced-dimension MUSIC [J]. IEEE Communications Letters,2010,14(12):1161 - 1163.

[163] 李建峰,张小飞,汪飞.基于四元数的 Root-MUSIC 的双基地 MIMO 雷达中角度估计算法 [J].电子与信息学报,2012,34(02):300 - 304.

[164] ZHANG X ,XU D.Angle estimation in bistatic MIMO radar using improved reduced dimension Capon algorithm [J]. Journal of Systems Engineering and Electronics,2013,24(1):84 - 89.

[165] ZHANG X, CHEN C, LI J. Angle estimation using quaternion-ESPRIT in bistatic MIMO radar [J]. Wireless Personal Communications,2013,69(2):551 - 560.

[166] CHEN C,ZHANG X,CHEN H,et al.A low-complexity algorithm for coherent DOA estimation in monostatic MIMO radar [J]. Wireless Personal Communications,2013,72:549 - 563.

[167] LI J,ZHANG X,CAO R,et al.Reduced-dimension MUSIC for angle and array gain-phase error estimation in bistaticMIMO radar [J]. IEEE Communications Letters,2013,17(3):443 - 446.

[168] LI J,ZHANG X.Closed-form blind 2D-DOD and 2D-DOA estimation for MIMO radar with arbitrary arrays [J]. Wireless Personal Communications,2013,69:175 - 186.

[169] LI J,ZHANG X.Unitary subspace-based method for angle estimation in bistatic MIMO radar[J].Circuits Systems and Signal Processing,2014,33(2):501 - 513.

[170] CHEN J,GU H,SU W.Angle estimation using ESPRIT without pairing in MIMO radar [J].Electronics Letters,2008,44(24):1422 - 1423.

[171] 陈金立,顾红,苏为民.一种双基地 MIMO 雷达快速多目标定位方法 [J].电子与信息学报,2009,31(7):1664 - 1668.

[172] 程院兵,顾红,苏卫民.一种新的双基地 MIMO 雷达快速多目标定位算法 [J].电子与信息学报,2012,34(2):312 - 317.

[173] CHEN J,GU H,SU W M.A new method for joint DOD and DOA estimation in bistatic MIMO radar [J].Signal Processing,2010,90

(2):714 - 718.

[174] CHENG Y, YU R, GU H, et al. Multi-SVD based subspace estimation to improve angle estimation accuracy in bistatic MIMO radar[J]. Signal Processing, 2013, 93(7): 2003 - 2009.

[175] 王伟,王咸鹏,李欣.MIMO 雷达参数估计技术[M].北京:国防工业出版社,2017.

[176] WANG X P, WANG W, LI X, et al. Sparsity-aware DOA estimation scheme for noncircular source in MIMO radar[J]. Sensors, 2016, 16(4): 539.

[177] LIU Q, WANG X P. Direction of arrival estimation via reweighted l_1 norm penalty algorithm for monostatic MIMO radar[J]. Multidimensional Systems and Signal Processing, 2018, 29(2): 733 - 744.

[178] LIU J, WANG X P, ZHOU W D. Covariance vector sparsity-aware DOA estimation for monostatic MIMO radar with unknown mutual coupling[J]. Signal Processing, 2016, 119: 21 - 27.

[179] WANGX P, WANG W, LI X, et al. Real-valued covariance vector sparsity-inducing DOA estimation for monostatic MIMO radar[J]. Sensors, 2015, 15(11): 28271 - 28286.

[180] WANG X P, WANG W, LIU J, et al. A sparse representation scheme for angle estimation in monostatic MIMO radar[J]. Signal Processing, 2015, 104: 258-263.

[181] WANG X P, WANG W, LI X, et al. A tensor-based subspace approach for bistaticMIMO radar in spatial colored noise[J]. Sensors, 2014, 14(3): 3897 - 3907.

[182] WANG X P, WANG W, LIU J, et al. Tensor-based real-valued subspace approach for angle estimation in bistatic MIMO radar with unknown mutual coupling[J]. Signal Processing, 2015, 116: 152 - 158.

[183] WANG W, WANG X P, LI X, et al. Conjugate ESPRIT for DOA estimation in monostatic MIMO radar[J]. Signal Processing, 2013, 93(7): 2070 - 2075.

[184] WANG X P, WANG W, XU D J. Low complexity ESPRIT-ROOT - MUSIC algorithm for non circular source in bistatic MIMO radar[J]. Circuits Systems and Signal Processing, 2015, 34(4): 1265 - 1278.

[185] WANG W, WANG X P, LI X, et al. DOA estimation for monostatic MIMO radar based on Unitary Root-MUSIC[J]. International Journal of electronics, 2013, 100(11): 1499 - 1509.

[186] WANG X P, WANG W, XU D J, et al. Matrix pencil method forbistatic MIMO radar with single snapshot[J]. IEICE transactions on electronics, 2014, 97(2): 120 - 122.

[187] WANG W, WANG X P, MA Y H, et al. Conjugate unitary ESPRIT algorithm for bistatic MIMO radar [J]. IEICE transactions on electronics, 2013, 96(1): 124 - 126.

[188] WANG W, WANG X P, LI X. Low-complexity method for angle estimation in MIMO radar [J]. IEICE transactions on communications, 2012, 95(9): 2976 - 2978.

[189] LIAO B. Fast angle estimation for MIMO radar with nonorthogonal waveforms[J]. IEEE Transactions on Aerospace and Electronic Systems, 2018, 54(4): 2091 - 2096.

[190] FRANKIE C, SO H C, LEI H. Parameter estimation and identifiability in bistatic multiple-input multiple-output radar [J]. IEEE Transactions on Aerospace and Electronic Systems, 2015, 51 (3): 2047 - 2056.

[191] HUANG L D, WANG X P, HUANG M X, et al. An implementation scheme of range and angular measurements for FMCW MIMO Radar via Sparse Spectrum Fitting[J]. Electronics, 2020, 9(3): 389.

[192] LIU F L, WANG X P, HUANG M X, et al. A novel unitary ESPRIT algorithm for monostatic FDA-MIMO radar [J]. Sensors, 2020, 20 (3): 827.

[193] MENG D D, WANG X P, HUANG M X, et al. Robust weighted subspace fitting for DOA estimation via block sparse recovery[J]. IEEE Communications Letters, 2019, 24(3): 563 - 567.

[194] WANG X P, ZHU Y H, HUANG M X, et al. Unitary matrix completion-based DOA estimation of noncircular signals in nonuniform noise[J]. IEEE Access, 2019, 7: 73719 - 73728.

[195] MENG D D, WANG X P, WAN L T, et al. Block rank sparsity-aware DOA estimation with large-scale arrays in the presence of unknown

mutual coupling[J].Digital Signal Processing,2019,94(4):96－104.

[196] GUO Y H,WANG X P,WAN L T,et al.Tensor-based angle and array gain-phase error estimation scheme in bistatic MIMO radar[J].IEEE Access,2019,7:47972－47981.

[197] WANG H F,WAN L T,DONG M X,et al.Assistant vehicle localization based on three collaborative base stations via SBL-based robust DOA estimation[J].IEEE Internet of Things Journal,2019,6(3):5766－5777.

[198] WANG X P,HUANG M X,SHEN C,et al.Robust vehicle localization exploiting two based stations cooperation:A MIMO radar perspective[J].IEEE Access,2018,6:48747－48755.

[199] WANG J X,WANG X P,XU D J,et al.Robust angle estimation for MIMO radar with the coexistence of mutual coupling and colored noise[J].Sensors,2018,18(3):832.

[200] WANG X P,HUANG M X,WU D J,et al.Direction of arrival estimation for MIMO radar via unitary nuclear norm minimization[J].Sensors,2017,17(4):939.

[201] WANG X P,WANG L Y,LI X M,et al.Nuclear norm minimization framework for DOA estimation in MIMO radar[J].Signal Processing,2017,135:147－152.

[202] GUO Y D,ZHANG Y S,TONG N N.Beamspace ESPRIT algorithm for bistatic MIMO radar[J].Electronics Letters,2011,47(15):876－878.

[203] GUO Y D,ZHANG Y S,TONG N N.ESPRIT-like angle estimation for bistatic MIMO radar with gain and phase uncertainties[J].Electronics Letters,2011,47(17):996－997.

[204] GUO Y D,ZHANG Y S,TONG N N.Central angle estimation of coherently distributed targets for bistatic MIMO radar[J].Electronics Letters,2011,47(7):462－463.

[205] 郭艺夺,张永顺,张林让,等.双基地 MIMO 雷达相干分布式目标快速角度估计算法[J].电子与信息学报,2011,33(7):1684－1688.

[206] 郭艺夺,张永顺,张林让,等.双基地 MIMO 雷达收发阵列互耦条件下目标定位方法[J].西安电子科技大学学报,2011,38(6):94－101.

[207] ZHENG G M,CHEN B X,YANG M G.Unitary ESPRIT algorithm

for bistatic MIMO radar [J].IET Electronics Letters,2012,48(3):179－181.

[208] ZHENG G M, TANG J, YANG X. ESPRIT and unitary ESPRIT algorithms for coexistence of circular and noncircular signals in bistatic MIMO radar [J].IEEE Access,2016,4(99):7232－7240.

[209] ZHENG G M. DOA estimation in MIMO radar with non-perfectly orthogonal waveforms [J]. IEEE Communications Letters, 2017, 21(2):414－417.

[210] ZHENG G M, CHEN B X. Unitary dual-resolution ESPRIT for joint DOD and DOA estimation in bistatic MIMO radar [J]. Multidimensional Systems and Signal Processing, 2015, 26(1):159－178.

[211] ZHENG G M. Beamspace Root-MUSIC algorithm for joint DOD DOA estimation in bistatic MIMO radar [J]. Wireless Personal Communications,2014,75(4):1879－1889.

[212] ZHENG G M, TANG J. DOD and DOA estimation in bistatic MIMO radar for nested and coprime array with closed-form DOF [J]. International Journal of Electronics,2017,104(5):885－897.

[213] SONG Y W, ZHENG G M, HU G P. A combined ESPRIT-MUSIC method for FDA-MIMO radar with extended range ambiguity using staggered frequency increment [J]. International Journal of Antennas and Propagation,2019,2019(1):1－7.

[214] SONG Y W, HU G P, ZHENG G M. Direction finding of coherent sources using a MIMO array of triaxial velocity sensors [J]. Mathematical Problems in Engineering,2019,2019:7010389.

[215] SHI J P, HU G P, ZONG B F, et al. DOA estimation using multipath echo power for MIMO radar in low-grazing angle[J]. IEEE Sensors Journal,2016,16(15):6087－6094.

[216] SHI J P, HU G P, LEI T. DOA estimation algorithms for low-angle targets with MIMO radar[J]. Electronics Letters,2016,52(8):652－654.

[217] SHI J P, HU G P, ZHANG X F, et al. Sparsity-based two-dimensional DOA estimation for coprime array: from sum-difference coarray viewpoint [J]. IEEE Transactions on Signal Processing,2017,65(21):5591－5604.

[218] CHEN P, CAO Z X, CHEN Z M, et al. Off-grid DOA estimation using sparse bayesian learning in MIMO radar with unknown mutual coupling[J]. IEEE Transactions on Signal Processing, 2019, 67(1): 208 - 220.

[219] WEN F Q. Computationally efficient DOA estimation algorithm for MIMO radar with imperfect waveforms[J]. IEEE Communications Letters, 2019, 23(6): 1037 - 1040.

[220] JIANG H, ZHANG J K, WONG K M. Joint DOD and DOA estimation for bistatic MIMO radar in unknown correlated noise[J]. IEEE Transactions on Vehicular Technology, 2015, 64(11): 5113 - 5125.

[221] ANTONIK P, WICKS M C, GRIFFITHS H D, et al. Frequency diverse array radars[C]// In Proc. IEEE Radar Conf. Dig., Verona, NY, USA, Apr.24-27, 2006: 215 - 217.

[222] WANG W Q. Frequency diverse array antenna: new opportunities[J]. IEEE Antennas and Propagation Magazine, 2015, 57(2): 145 - 152.

[223] SAMMARTINO P F, BAKER C J, GRIFFITHS H D. Frequency diverse MIMO techniques for radar [J]. IEEE Trans. Aerosp. Electron. Syst., 2013, 49(1): 201 - 222.

[224] WANG W Q, SO H C. Transmit subaperturing for range and angle estimation in frequency diverse array radar[J]. IEEE Trans. Signal Process., 2014, 62(8): 2000 - 2011.

[225] WANG W Q, SHAO H Z. Range-angle localization of targets by a double-pulse frequency diverse array radar [J]. IEEE J. Sel. Topics Signal Process., 2014, 8(1): 1 - 9.

[226] XU J W, LIAO G S, ZHU S Q, et al. Joint range and angle estimation using MIMO radar with frequency diverse array[J]. IEEE Trans. Signal Process., 2015, 63(13): 3396 - 3410.

[227] CHINTAGUNTA S, PONNUSAMY P. Integrated polarisation and diversity smoothing algorithm for DOD and DOA estimation of coherent targets[J]. IET Signal Processing, 2018, 12(4): 447 - 453.

[228] BENCHEIKH M L, WANG Y. Combined ESPRIT-Root MUSIC for DOD-DOA estimation in polarimetric bistatic MIMO radar [J]. Progress in Electromagnetics Research Letters, 2011, 22: 109 - 117.

[229] JIANG H, WANG D, LIU C. Joint parameter estimation of DOD DOA/polarization for bistatic MIMO radar[J]. The Journal of China Universities of Posts and Telecommunications, 2010, 17(5): 32 - 37.

[230] JIANG H, ZHANG Y, LI J, et al. A PARAFAC-based algorithm for multidimensional parameter estimation in polarimetric bistatic MIMO radar [J]. EURASIP Journal on Advances in Signal Processing, 2013, 2013(1): 1 - 14.

[231] 王克让,朱晓华,何劲.基于矢量传感器 MIMO 雷达的 DOD DOA 和极化联合估计算法[J].电子与信息学报,2012,34(1):160 - 165.

[232] 王克让,何劲,贺亚鹏,等.基于矢量传感器的扩展孔径双基地 MIMO 雷达多目标定位算法[J].电子与信息学报,2012,34(4):582 - 586.

[233] GU C, HE J, LI H, et al. Target localization using MIMO electromagnetic vector array systems[J]. Signal Processing, 2013, 93(7): 2103 - 2107.

[234] 樊劲宇,顾红,苏卫民,等.偶极子分离的矢量阵 MIMO 雷达多维角度估计算法[J].电子与信息学报,2013,35(8):1841 - 1846.

[235] WEN F Q, SHI J, ZHANG Z. Joint 2D-DOD, 2D-DOA and polarization angles estimation for bistatic EMVS-MIMO radar via PARAFAC analysis [J]. IEEE Transactions on Vehicular Technology, 2020, 29(2): 1626 - 1638.

[236] WEN F Q, SHI J. Fast direction finding for bistatic EMVS-MIMO radar without pairing[J]. Signal Processing, 2020, 173: 107512.

[237] WANG X P, WAN L T, HUANG M X, et al. Polarization channel estimation for circular and non-circular signals in massive MIMO systems[J]. IEEE Journal of Selected Topics in Signal Processing, 2019, 13(5): 1001 - 1016.

[238] MA H H, TAO H H, SU J, et al. DOD/DOA and polarization estimation in MIMO systems with spatially spread dipole quints[J]. IEEE Communications Letters, 2020, 24(1): 99 - 102.

[239] 郑桂妹,杨明磊,陈伯孝,等.干涉式矢量传感器 MIMO 雷达的 DOD/DOA 和极化联合估计[J].电子与信息学报,2012,34(11):2635 - 2641.

[240] 郑桂妹,陈伯孝,杨明磊.基于矢量传感器 MIMO 雷达的发射极化优化 DOA 估计算法 [J].电子与信息学报,2014,36(3):565 - 570.

[241] ZHENG G M, WU B. Polarisation smoothing for coherent source direction finding with MIMO electromagnetic vector sensor array [J]. IET Signal Processing, 2016, 10(8): 873 - 879.

[242] ZHENG G M, TANG J. Two-dimensional DOA estimation for monostatic MIMO radar with electromagnetic vector received sensors [J]. International Journal of Antennas and Propagation, 2016, 2016(2): 1 - 10.

[243] ZHENG G M, ZHANG D. BOMP-based angle estimation with polarimetric MIMO radar with spatially spread crossed-dipole [J]. IET Signal Processing, 2018, 12(1): 113 - 118.

[244] NEHORAI A, PATAN E. Acoustic vector-sensor array processing [J]. IEEE Trans. Signal Process., 1994, 42(9): 2481 - 2491.

[245] HAWKES M, NEHORAI A. Acoustic vector-sensor processing in the presence of a reflecting boundary [J]. IEEE Trans. Signal Process., 2000, 48(11): 2981 - 2993.

[246] HE J, SWAMY M N S, AHMAD M O. Joint DOD and DOA estimation for MIMO array with velocity receive sensors [J]. IEEE Signal Process. Lett., 2011, 18(7): 399 - 402.

[247] LI J, ZHANG J. Improved joint DOD and DOA estimation for MIMO array with velocity receive sensors [J]. IEEE Signal Process. Lett., 2011, 18(12): 717 - 720.

[248] LI J, ZHANG X. Two-dimensional angle estimation for monostatic MIMO arbitrary array with velocity receive sensors and unknown locations [J]. Digit. Signal Process., 2014, 24(24): 34 - 41.

[249] TAO J, CHANG W, CUI W. Vector field smoothing for DOA estimation of coherent underwater acoustic signals in presence of a reflecting boundary [J]. IEEE Sensors J., 2007, 7(8): 1152 - 1158.

[250] TAO J, CHANG W, SHI Y. Direction-finding of coherent sources via "particle-velocity-field smoothing" [J]. IET Radar Sonar Navig., 2008, 2(2): 127 - 134.

[251] HE J, JIANG S, WANG J, et al. Particle-velocity-field difference smoothing for coherent source localization in spatially nonuniform noise [J]. IEEE J. Oceanic Eng., 2010, 35(1): 113 - 119.

[252] HEINER K. VHF/UHF radar part 1: characteristics[J]. Electron. Commun. Eng. J., 2002, 14(2): 61-72.

[253] HEINER K. VHF/UHF radar part 2: operational aspects and applications[J]. Electron. Commun. Eng. J., 2002, 14(3): 101-111.

[254] LIU Y, LIU H, XIA X G, et al. Projection techniques for altitude estimation over complex multipath condition-based VHF radar [J]. IEEE J. Sel. Topics Appl. Earth Obs. Remote Sens., 2018, 11(7): 2362-2374.

[255] 王胜华.雷达低仰角目标检测与测高关键技术研究[D].西安:西安电子科技大学,2019.

[256] 谭俊.米波雷达低仰角测角中多径效应影响抑制及关键技术研究[D].成都:电子科技大学,2019.

[257] XIANG H, CHEN B, YANG M, et. al. Altitude measurement based on characteristics reversal by deep neural network for VHF radar [J]. IET Radar Sonar Navig., 2019, 13(1): 98-103.

[258] 吴向东,赵永波,张守宏,等.一种 MIMO 雷达低角跟踪环境下的 DOA 估计新方法[J].西安电子科技大学学报,2008,35(5):793-798.

[259] ELDAR Y C, KUPPINGER P, BOLCSKEI H. Block-sparse signals: Uncertainty relations and efficient recovery [J]. IEEE Trans. Signal Process., 2010, 58(6): 3042-3054.

[260] 王永良,陈辉,彭应宁,等.空间谱估计理论与方法[M].北京:清华大学出版社,2004.

[261] CHEN C, TAO F, ZHENG G M, et al. Beam split algorithm for height measurement with meter-wave MIMO radar [J]. IEEE Access, 2021, 9: 5000-5010.

[262] TAN J, NIE Z, PENG S. Adaptive time reversal MUSIC algorithm with monostatic MIMO radar for low angle estimation [C]//2019 IEEE Radar Conference (RadarConf), Boston, MA, USA, 2019: 1-6.

[263] HARRINGTON R F. Time harmonic electromagnetic fields [M]. New York: McGraw Hill, 1961.

[264] TAN J, NIE Z P. Polarisation smoothing generalised MUSIC algorithm with PSA monostatic MIMO radar for low angle estimation [J]. Electronics Letters, 2018, 54(8): 527-529.

[265] YUAN L, HONG W L, XIANG G X, et al. Projection techniques for altitude estimation over complex multipath condition-based VHF radar [J]. IEEE Journal of Selected Topics in Applied Earth Observations & Remote Sensing, 2018, 11(7): 2362 - 2375.